# Drying of Aromatic Plant Material for Natural Perfumes

*Drying of Aromatic Plant Material for Natural Perfumes* provides readers with a deep understanding of the fascinating world of aromatic plants, drying, extraction and perfumery. It covers the significance and global demand of essential oils, a glimpse of plant histology, secretion and storage of secondary metabolites in plant tissues, drying technologies and selection for dehydration of herbage, extraction, chemical compositions and applications. The book is organized into four sections: plant cell structures and their role during dehydration, aromatic plants and their essential oil contents, composition and contribution towards perfumery, and dehydration and extraction technologies. The book does the following:

- Summarizes commercially important aromatic plant materials and extracted essential oil around the world and their socio-commercial impact
- Includes structures of plant tissues, inherent mass transfer pathways, along with the presence and role of water, the understanding of which can aid dehydration and extraction
- Describes methods for selecting drying parameters and relates drying mechanisms with microstructure
- Discusses drying technologies for heat-sensitive materials and extraction technologies for selective component extractions
- Covers preprocessing, extraction and standardization of essential oils for blending for different applications

This book serves as a handy tool for R&D, industrial, and academic researchers working in perfumery, fragrance, and food science, as well as chemical and agricultural engineering.

# Advances in Drying Science and Technology
Series Editor Arun S. Mujumdar
*McGill University, Quebec, Canada*

*Frontiers in Spray Drying*
Nan Fu, Jie Xiao, Meng Wai Woo, and Xiao Dong Chen

*Drying in the Dairy Industry*
Cécile Le Floch-Fouere, Pierre Schuck, Gaëlle Tanguy, Luca Lanotte, and Romain Jeantet

*Spray Drying Encapsulation of Bioactive Materials*
Seid Mahdi Jafari and Ali Rashidinejad

*Flame Spray Drying: Equipment, Mechanism, and Perspectives*
Mariia Sobulska and Ireneusz Zbicinski

*Advanced Micro-Level Experimental Techniques for Food Drying and Processing Applications*
Azharul Karim, Sabrina Fawzia, and Mohammad Mahbubur Rahman

*Mass Transfer Driven Evaporation of Capillary Porous Media*
Rui Wu and Marc Prat

*Particulate Drying: Techniques and Industry Applications*
Jangam Vinayak, Chung-Lim Law, and Shivanand Shirkole

*Drying and Valorisation of Food Processing Waste*
Chien Hwa Chong, Rafeah Wahi, Chee Ming Choo, Shee Jia Chew, and Mackingsley Kushan Dassanayake

*Drying of Herbs, Spices, and Medicinal Plants*
Ching Lik Hii and Shivanand Shirkole

*Drying of Aromatic Plant Material for Natural Perfumes*
Viplav Hari Pise, Ramakant Harlalka, and Bhaskar Narayan Thorat

For more information about this series, please visit: www.routledge.com/Advances-in-Drying-Science-and-Technology/book-series/CRCADVSCITEC

# Drying of Aromatic Plant Material for Natural Perfumes

Viplav Hari Pise
Ramakant Harlalka
Bhaskar Narayan Thorat

CRC Press
Taylor & Francis Group
Boca Raton London New York

CRC Press is an imprint of the
Taylor & Francis Group, an **informa** business

First edition published 2024
by CRC Press
2385 Executive Center Drive, Suite 320, Boca Raton, FL 33431

and by CRC Press
4 Park Square, Milton Park, Abingdon, Oxon, OX14 4RN

*CRC Press is an imprint of Taylor & Francis Group, LLC*

ISBN: 978-1-032-32502-6 (hbk)
ISBN: 978-1-032-32504-0 (pbk)
ISBN: 978-1-003-31538-4 (ebk)

DOI: 10.1201/9781003315384

Typeset in Times New Roman
by MPS Limited, Dehradun

# Contents

# Advances in Drying Science and Technology

*Series Editor Dr. Arun S. Mujumdar*

It is well known that the unit operation of drying is a highly energy-intensive operation encountered in diverse industrial sectors, ranging from agricultural processing, to ceramics, chemicals, minerals processing, pulp and paper, pharmaceuticals, coal polymer, food, forest products industries as well as waste management. Drying also determines the quality of the final dried products. The need to make drying technologies sustainable and cost effective via application of modern scientific techniques is the goal of academic as well as industrial R&D activities around the world.

Drying is a truly multi- and interdisciplinary area. Over the last four decades the scientific and technical literature on drying has seen exponential growth. The continuously rising interest in this field is also evident from the success of numerous international conferences devoted to drying science and technology.

The establishment of this new series of books entitled *Advances in Drying Science and Technology* is designed to provide authoritative and critical reviews and monographs focusing on current developments as well as future needs. It is expected that books in this series will be valuable to academic researchers as well as industry personnel involved in any aspect of drying and dewatering.

The series will also encompass themes and topics closely associated with drying operations, e.g., mechanical dewatering, energy savings in drying, environmental aspects, life cycle analysis, technoeconomics of drying, electrotechnologies, control and safety aspects, and so on.

## ABOUT THE SERIES EDITOR

**Dr. Arun S. Mujumdar** is an internationally acclaimed expert in drying science and technologies. He was the founding chair in 1978 of the International Drying Symposium (IDS) series and has been editor-in-chief of *Drying Technology: An International Journal* since 1988. The fourth enhanced edition of his *Handbook of Industrial Drying,* published by CRC Press, has just appeared. He is the recipient of numerous international awards, including honorary doctorates from Lodz Technical University, Poland, and University of Lyon, France.

Please visit https://arunmujumdar.com/

# Authors

**Viplav Hari Pise** is Prof M. M. Sharma Research Fellow from the Department of Chemical Engineering, Institute of Chemical Technology Mumbai, India, where he also pursued his PhD. He has more than 6 years of industrial experience as Research & Process engineer at Larsen & Toubro Hydrocarbon Engineering Limited. With experience in project execution and management he pursues research in value addition to natural products through extraction and isolation of phytochemicals. His broad areas of research are identification of indigenous aromatic and medicinal plants, developing extraction protocols, scale-up for obtaining isolates for market acceptance and commercialization of the project.

**Ramakant Harlalka** is the founder and director of Nishant Aromas and has more than 35 years of experience working with essential oils and their applications, as well as extensive experience with large-scale aromatic plantation, extraction, blending and formulation and marketing. He is the founder of multiple business ventures and is a prestigious member of various research councils and industry associations.

**Bhaskar Narayan Thorat** is the founder and director of ICT–IOC, Bhubaneswar, and professor of Chemical Engineering, Institute of Chemical Technology, Mumbai, India. He has over 30 years of experience in developing grassroots and sustainable technologies for the welfare of society. He has vast experience in research and development in drying, crystallization and filtration and has conducted research on dehydration of agri-produce and marine resources through multiple modes of drying and developed patented technologies suitable for field applications. He initiated the World Forum for Crystallization, Filtration and Drying (WFCFD) in 2006.

# 1 Aromatic Plants – Significance and Impacts

## 1.1 SIGNIFICANCE OF AROMATIC PLANTS

Plants are known for their ornamental value and providing oxygen and food in the food chain. Plants are a significant source of raw materials for the non-food industrial sector, including *primary metabolites* (oils, carbohydrates and fibres or even biomass) in bulk quantities to industries and *secondary metabolites*, albeit in small quantities, used for the production of speciality products like essential oils (EOs), pharmaceuticals, herbal products, natural dyes and colourants, and so on.

Plants mainly synthesize two kinds of oil: fixed oils and EOs. *Fixed oils* consist of esters of glycerol (triglycerides or triacylglycerols) and fatty acids. These are required in the human diet to maintain good physical and mental health. *EOs* are complex mixtures of highly concentrated, volatile, and hydrophobic compounds. They contribute to a plant's characteristic flavour and fragrance (F&F). The term *essential oil* dates back to the sixteenth century, stating it to be merely *a pure scented oil* or *volatile scented oil of natural biological origin.* "Essential" reflects the plant's essence or intrinsic nature, and "oil" refers to insolubility in water. Plants secrete metabolites for functions such as protection (from competitors, pathogens or insects, and so on), reduction in transpiration losses, communication with other plants and microbes, and attraction of the pollinating agents. These metabolites consist of volatile and non-volatile compounds (Firn, 2010).

As per the International Organisation of Standards, ISO 9235:2013, EOs are defined as "*Product obtained from natural raw materials of vegetal origin, by steam distillation, or dry distillation, after the separation of the aqueous phase – if any – by physical process. In the specific case of citrus fruits, the oil is obtained by pressing at room temperature of the epicarp – cold-pressed essential oil (EOs).*" Agence Française de Normalisation (AFNOR) gives the following definition (NF T 75–006): "*The essential oil is the product obtained from a vegetable raw material, either by steam distillation or by mechanical processes from the epicarp of Citrus, or 'dry' distillation. The essential oil is then separated from the aqueous phase by physical means*" (Dhifi et al., 2016). EOs, as defined by the *European Pharmacopeia, Seventh Edition,* are "*Odorant products, which have the complex composition, and obtained from raw plant extract, either extracted by the steam of water, dry distillation or a suitable mechanical method without heating. Generally, a physical method is used to separate the essential oil from the aqueous phase, which has no significant change in its chemical composition.*"

*EOs* are secondary metabolites contributing to odoriferous constituents or the essence of aromatic plants. These are known to be secreted directly by protoplasm or hydrolysis of glycosides and are associated with plant structures of glandular

DOI: 10.1201/9781003315384-1

hairs in the *Lamiaceae* family, oil tubes or vittae in the *Apiaceae* family, modified parenchymal cells in the *Piperaceae* family, Schizogenous or lysigenum passages in the *Rutaceae* family. EOs are obtained from different parts of plants like bark, petioles, leaves, seeds, stems, flowers and flower parts, fruits, roots or rhizomes, secreting and storing these compound compositions. Single-cell or multicellular trichomes, both glandular and non-glandular (depending on the morphology and secretion ability), are seen in most plant species. These trichomes are known for the key and unique feature of secreting a number of specialized metabolites (Huchelmann et al., 2017). The pearl glands are observed to store these volatiles and can be preserved in the natural matrix for further extraction (Boukhris et al., 2013; Pise et al., 2022; Rehman & Asif Hanif, 2016). These secondary metabolites, both volatiles and EOs, have a diverse array of properties and can be used for therapeutic actions and positive effects on health and wellbeing. The size and complexity of the stereochemistry make it extremely complicated to track and study the occurrence or separation. However, the significance of these metabolites, of which EOs are a part, has sought the attention of various researchers and resulted in reporting several studies on the topic, which are further compiled in this book.

## 1.2   UNIT OPERATIONS – DEHYDRATION AND EXTRACTION

The extraction of bioproducts means separating the complex mixtures of secondary metabolites from the natural matrix, as stated earlier (Sharifi-Rad et al., 2017). This extraction process can be simplified as a series of unit operations on a laboratory or industrial scale. These include the mass transfer of the plant volatiles from the biomass to the solvent/carrier fluid, separation of the solvent/carrier fluid and the biomass, separation of the plant volatiles from the solvent/carrier fluid and purification of the essential/volatile oils. Also, removing existing water in the herbage is crucial to lower the utility consumption (for water/steam distillation) and minimize water interference with other solvents (for solvent extraction). This mass transfer operation, specific to water molecules without the impact on the bioproducts, makes drying a critical function. Though well-reported, these mass transfer operations of drying, extraction/leaching, and separation hold a special consideration in the case of NVEO extraction post-harvesting of aromatic plants owing to the low shelf life and tender structure of plants and localized low concentration of desired secondary volatiles.

Drying at the molecular level involves activating water molecules, mobilizing them within the matrix, and transferring them outside the matrix. The product quality is closely related to the way the dehydration process is carried out. The dehydration process should retain the product's characteristics from the application point of view (Thamkaew et al., 2021). The most desired purpose of dehydration is to reduce moisture in herbs, spices or other parts of aromatic and medicinal plants without affecting their key attributes for further use (Bhaskara Rao & Murugan, 2021). Dehydration takes place by applying heat and mass transfer at the cellular level. The quantification and ways of removing free-water, inter-, and intra-cellular water must be understood through experimental and mathematical analysis (Khan et al., 2017). The detailed analysis regarding the dehydration at the cellular level is not only valuable for determining the parameters affecting the drying, including activation energy, driving force/concentration gradient, internal & external mass transfer rate and effective moisture diffusion (Majumder et al., 2021), but also it will help in the understanding of the conditions required for the retention of desired phytochemicals and volatiles in herbs, spices and aromatic plants (Pise et al., 2022; Prothon et al., 2003). It is also highly desired to obtain the critical parameters for energy conservation vis-à-vis the desirable thermal conditions (Rahman et al., 2018). These parameters can be controlled by mode of energy application, temperature, flow/draft & relative humidity of the drying medium and size of the material being dried. The advances in drying technology through different dehydrators and solar dryers for agrocommodities, food applications (Calín-Sánchez et al., 2020; Radoj et al., 2021; Uthpala et al., 2020), and aromatic plants (Bhaskara Rao & Murugan, 2021; Jin et al., 2018; Majumder et al., 2021; Orphanides et al., 2016; Qiu et al., 2020; Thamkaew et al., 2021) have been well reported.

In the case of extraction, the first unit operation of mass transfer can be carried out by different extraction methodologies using fluids like water/steam, organic solvents (polar or non-polar), critical fluids (sub- and super-), or fats and oils, or by mechanical means, with or without temperature, depending on the final applications. The composition of a complex compound mixture extracted, as plant volatiles (and non-volatiles in some cases), depends on various conditions at which the extraction is performed and partially on the extraction method. The extraction process can also be subjected to specific procedures and parameters for selective separation (Zhang et al., 2018). Opting for appropriate extraction methods, followed by necessary separation, isolation, and purification for obtaining volatile oils of commercial interest, is important. Similarly, for getting EOs, hydro-, steam- or dry-distillation, or mechanical process without heating (termed "expression," especially for citrus oils) as per the definitions is to have opted.

Bioproducts, such as these volatiles, need assurance of constancy and quality to ensure safe and efficient operations for industrial applications, which is difficult to achieve. As mentioned above, the constituent of these volatiles varies with various factors, including extraction parameters (Figueiredo, 2017; Tisserand & Young, 2014). However, by limiting the variables like specific region, fixed maturity of harvests, same extraction process and uniform conditions, orderly pre-processing and so on, variation in the composition can be minimized. The combined effects of the constituents extracted lend to the oil characteristics such as odour and therapeutic properties. Hence, a thorough understanding of the extraction process, the impact on the extract's composition, and the reason for post-extraction applications are critical for determining the post-harvest unit operation of the drying and extraction process and obtaining the desired utilizable volatile oils.

## 1.3 GEOGRAPHY, KEY MARKETS AND CURRENT TRENDS

It is difficult to obtain the exact data on the EO-bearing plants and their production. There are about 350,000 estimated known plant species globally, and around 17,500 species (approx. 5%) are aromatic. However, only about 400–500 are

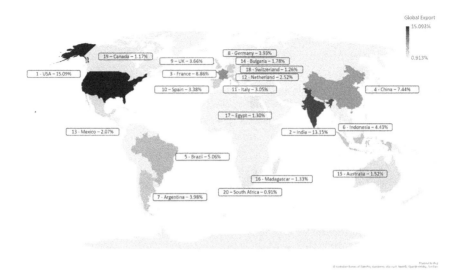

**FIGURE 1.1**    Top 20 Essential oil exporting countries across the globe.

commercially processed as aromatic raw materials, and hardly 50% are cultivated (Tisserand & Young, 2014). The main types of EOs are orange oil, lemon oil, lime oil, peppermint oil, corn mint oil, citronella oil, spearmint oil, geranium oil, clove leaf oil, and eucalyptus oil. The world production and consumption of EOs and natural volatiles are increasing exponentially. The produced EOs are amongst the top 500 most traded products, with a reported total trade of about $6.31 billion in 2019, $5.41 billion in 2020, $8.8 billion in 2021 and $9.62 billion in 2022. The expected growth of the EOs market globally is expected to be $18.25 billion in 2028 at a compound annual growth rate (CAGR) of 9.57%. The top exporters of EOs were the United States ($816 M), India ($712 M), France ($480 M), China ($403 M), and Brazil ($274 M) (Figure 1.1). The top importers of EOs were the United States ($1.05B), France ($414 M), China ($364 M), Germany ($350 M), and the Netherlands ($295 M) (Simoes & Hidalgo, 2021) (Figure 1.2). The estimated production of EOs in 2004 was around 45,000 tons of EOs and about 104,000 tons by 2009 (Lubbe & Verpoorte, 2011).

Across the globe, EOs are selectively used due to their functional and biological properties but are widely used as fragrance ingredients in perfumes, toiletries, detergents, food and beverages, textiles, and cosmetics. The contribution of EOs is about 55%–60% for flavours in the food industry, 15%–21% for fragrances in the perfumery/cosmetic industry, 10%–20% for isolation of components, 5%–10% as active substances in pharmaceutical preparations and 2%–5% for natural products (Joy, 2007). Though considered an industrial raw material, variability is seen in the type of extract, botanical certification, chemical polymorphism and assay conditions through various sources. Factors like genomics, edaphoclimatic variation, and seasonality influence the raw material quality and are much more important than processing (Figueiredo, 2017). In the case of stable and consistent requirements, selecting the right plantation material and processing

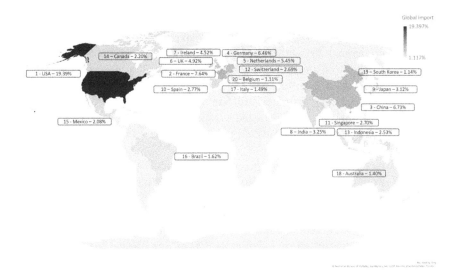

**FIGURE 1.2** Top 20 Essential oil importing countries across the globe.

technology can help meet the market demand and extract the compounds with improved overall yield and quality.

## 1.4 SOCIO-COMMERCIAL IMPACTS

It is mostly seen that remote rural areas or forest reserves are rich in natural resources. These areas suffer from different combinations of problems, including governance and market & resource endowment failures (Deshingkar & Akter, 2009). Well-planned cultivation of aromatic crops in such regions will not just provide a significant increase in production but also result in better utilization of the area and local employment.

Annual production of EOs is around 16,000 to 17,000 tons per year in India. Out of this, approximately 5,000 tons of oils are supplied as a raw material for perfumery with a valuation of about $48.5 million (INR 400 crores), and an export of around $15.8 million (INR 130 crores) is reported. Domestic demand for EO is fulfilled with indigenous production (about 90%), and the remaining is imported (Sanganeria, 2014).

The EO extraction industry across the world is a labour-intensive industry. In this section, the cultivation and extraction of EOs are reported in different countries, and the impact is seen on the local social life there.

### 1.4.1 PETITGRAIN OIL – PARAGUAY

Petitgrain oil, extracted from the leaves and twigs of the bitter orange tree (*Citrus aurantium L.*), yields about 0.30%–0.35% of oil with key components of linalyl acetate and linalool. Plant harvesting begins from the fifth year and lasts 35–40 years. Petitgrain oil has applications as a fragrance component for all kinds of perfumes, cosmetics and household chemicals. It is produced by steam-distilling harvested leaves and twigs (Gade, 1979). The harvest is taken twice a year: between October and February, around 70%–80% and in April and June, around 20%–30%.

Annual global consumption of this oil is reported to be around 250 tons. Paraguay is the main exporter, with 80% of global consumption. The main production departments in Paraguay are San Pedro (accounting for about 92% of production), Canindeyu, Cordillera and Caaguazu, accounting for a total production of 200 tons per year. Studies indicate 80% of the species in this region are cultivated, and 20% grow wild.

The extraction process is done manually without mechanization, from cutting the leaves to distillation. The distillation is carried out at farm level in wooden home stills with a capacity of 400 kg of leaves per batch. These farm distillation units are used by several farmers located nearby.

The entire production of 200 tons completely depends upon smallholders with an average land holding of about 1–2 hectares. There are about 15,000 families with livelihoods linked with the petitgrain oil business. Petitgrain oil improves with time if properly stored, can be extracted all year long and generates cash when there is no income from any other alternative product. Hence, it is considered important oil for farmers (Raul, 2020).

### 1.4.2   ROSE OIL – TURKEY AND BULGARIA

Rose oil is obtained from rose flowers as EO, absolute and concrete. The rose oil content in flowers ranges from 0.045% to 0.055%, with major components of citronellol, nerol, geraniol and demascone. There are more than 200 species in the *Rosa* genus; however, only four main species of *Rosa x damascena Mill., Rosa centifolia, Rosa alba and Rosa sertata/rosa rugosa* are commercially cultivated (Chalova et al., 2017). Rose oil and rose concrete are mainly consumed by big cosmetics and perfumery companies. The harvesting of rose flowers continues for less than a month, starting from the second half of May. Rose oil harvesting and extraction is a tough and labour-intensive activity carried out by families without high-income expectations because it is a traditional local product and a part of cultural heritage. In 2017, rose *damascena* petal prices were +/- $1.80 (INR 130–135) per kg, with the women pickers being paid $0.40 (INR 30–35) per kg for their labour.

The global consumption of rose oil accounts for 3,000 to 4,500 kg, 80%–90% of which is produced in Bulgaria and Turkey. The other key producers of rose oil are Morocco, Iran, Mexico, France, Italy, Lebanon, India, Russia, China and, to a small extent Afghanistan, Saudi Arabia and Egypt (Kovacheva et al., 2010).

#### 1.4.2.1   Bulgaria

Commercial cultivation is estimated on around 3,500–4,000 hectares of land. The short season of harvest and manual plucking of flowers makes it labour-intensive, involving around 40,000 pickers and more than 12,000 people working in the rose industry year-round. Around 7,000 families in mountainous and semi-mountainous areas earn their income from oil plant cultivation. Most extraction of oil is by significant distillers.

Hence, the rose industry in Bulgaria influences the livelihood of around 65,000–70,000 people. In 2017, the estimated production based upon a harvest of 11,000/12,000 MT of flowers was around 2,400 kg of oil, 2,000 kg of concrete and 200 kg of absolute worth around $30.2 million (Bleimann, 2019; Kovacheva et al., 2010).

### 1.4.2.2 Turkey

In 2010, rose cultivation was estimated from around 1,600 hectares of land involving some 10,000 families in oil, concrete and absolute production. Most rose flower production comes from small family plots of less than 1 hectare, which is supplied to one of six major distilleries producing about 65% of Turkey's EO. In 2013, some 7,000–8,000 MT of roses were harvested and sent to the distilleries producing about 1,400 kg of EO, 6,000 kg of rose concrete, and subsequently 1,000 kg of absolute. In 2017, annual production was estimated to be around 1,400 kg of EO, 10,000 kg of concrete and 5,000 kg of absolute, with an approximate worth of about $45.14 million(Bleimann, 2019; Giray & Omerci Kart, 2012).

### 1.4.2.3 Iran

It is reported that some 5,000 hectares are under cultivation in areas of Kashan, Kerman, Shiraz and Kermanshah. These farms are generally small land holdings producing low EO of only about 200 kg annually, but the production of rose water is about 3.85 million kg (90% of global rose water demand) worth $8 million (IFEAT Rose report, 2019).

### 1.4.2.4 Morocco

Around 880 hectares of land are cultivated, producing approximately 2,000 MT of flowers employing 6,000 small farmers majorly utilized for concrete production.

### 1.4.2.5 India

In India, 2,500–3,000 hectares of *Rosa damascena* are under cultivation, producing approximately 200 kg of oil and larger quantities of rose water.

### 1.4.2.6 Afghanistan

Afghanistan is an upcoming producer of rose products, with almost 3,000 hectares planted with rootstock and two or three distillation facilities brought from Bulgaria and Turkey. Rose cultivation is estimated to involve more than 400 farmers from the dangerous area of eastern Afghanistan (IFEAT Rose report, 2019).

### 1.4.3 JASMIN OIL – INDIA

More than 80 species of the *Jasminum* genus are found in India, of which four species are mainly cultivated for perfumery and fragrances: *Jasminum grandiflorum* and *Jasminum sambac*, *Jasminum asteroides* and *Jasminum auriculatum*.

Jasmine does not yield an oil by steam distillation, i.e. jasmine EO, as defined by ISO 3218. The commonly employed extraction method for almost 98% of jasmine production worldwide is a two-step process; solvent extraction as concrete followed by separation of wax and absolute. Jasmine concrete is extracted using hexane and then converted into a jasmine absolute by washing it with ethanol and separating wax at low temperatures.

The two main producers of *Jasminum grandiflorum* concrete are India and Egypt, accounting for 95% of the global market share. In 2014, the estimated concrete production in India was about 5.5–6 tons, whereas in Egypt it was about 4.5 tons. In India, the season lasts from June to December, with peak production in August/September for grandiflorum and from March to October for Sambac. In Egypt, the production period is typically from June to October but may be extended from end-May to early December.

### 1.4.3.1 India

Jasmin grandiflorum is cultivated throughout peninsular India, but in the Coimbatore district of Tamil Nadu, it is mainly cultivated under contract for extraction. The total cultivation area of grandiflorum in Tamil Nadu alone is about 2,850 hectares, of which only 10%–15% is processed for extraction. However, approximately 90% of grandiflorum flowers cultivated in the Coimbatore district are used for extraction. Grandiflorum has a productive life from the third to the tenth year, yielding 5–6 tons of flowers per hectare per year. The concrete yield from flowers ranges from 0.27% to 0.3%.

*Jasminum sambac* is also cultivated throughout peninsular India and, to a lesser extent, in the Gangetic plains. The total cultivation area of *Sambac* in Tamil Nadu is almost 6,500 hectares. The districts of Madurai, Virudhunagar, Theni, Dindigul and Sivaganga have been granted a Geographical Indication Mark for the *Jasminum sambac* flowers grown. *Sambac* has a productive life from the third to the eighth year with a yield of 4–5 tons of flowers per hectare per year, yielding concrete between 0.12% and 0.13%.

Considering an area of around 6,500 hectares under Jasmin cultivation in India, production involves around 20,000 small and marginal farmers. Around 80,000–100,000 people are involved in the picking activity, collecting an average of 3–4 kg of flowers for extraction. Usually, all family members work on this activity. A hectare of land can produce around 6 metric tons of flowers annually. The cost of cultivation average could be around $4,000 per year per hectare (including harvesting costs, which are the biggest ones) (IFEAT Jasmine report, 2015; Saripalle, 2016).

### 1.4.3.2 Egypt

Cultivation of *Jasmin grandiflorum* located in the Nile delta, around the village of Shoubra Beloula El-Sakhaweya (commune of Kutur, province of Gharbeya) accounts for 99% of jasmine plantations and the rest in the Fayoum area. An area varying between 105 and 150 hectares is under cultivation of jasmine, supporting approximately 5,000 flower pickers and 30,000 people linked through dependent family members and businesses. The Jasminum grandiflorum is processed mainly at the farm/district level. The plants' productive life is over 25 years, but they generally replace every 12–15 years of drop in acreage productivity. A single hectare of land can produce around 9.5–14.2 tons of blossoms per year, with a concrete yield of 0.26%–0.31%. This concrete yields about 55% and 61% of jasmine absolute and the rest as jasmine wax for cosmetics, candles and wood furnishing polish treatment applications.

Jasmine produce is almost completely exported, giving the country the highest consistent return worth $6.5 million in value (IFEAT Jasmine report, 2015).

### 1.4.4   LAVENDER – BULGARIA, UKRAINE, FRANCE AND SPAIN

Lavender oil, similar to rose, is extracted as EO through steam distillation as well as concrete and absolute through solvent extraction. Lavender originates in the Mediterranean basin, in rocky, calcareous areas of north Africa, Mediterranean countries, Europe and Western India. The flowers and leaves in smaller quantities are used for oil extraction. Out of 48 known species, only three are cultivated commercially to extract EOs – *L. angustifolia*, *L. latifolia*, *L. angustifolia x L. latifolia* (lavandin) (DAFF, 2009). The oil content in the dried lavender harvest is 2.1%–4%, with major components of Linalool, Linalyl acetate and lavandulyl acetate. With a global production of 1,300 tons of EO, major production is from Bulgaria and France, followed by Ukraine, Australia, Spain, England and Italy.

#### 1.4.4.1   Bulgaria

Almost 4,500 hectares of land are under lavender cultivation. Presently, with an average yield of 40 kg/ha of oil, about 150–180 tons are being produced from over 30 distilleries, with a capacity starting from 1–2 tons and extending to higher capacities. In 2013, around 1,200–1,300 farmers and 300–350 people were engaged in cultivation and oil production, respectively. In Bulgaria, around 8,000–10,000 people have their livelihood depending on the lavender oil business, including farming, processing, pickers, dependent families, agronomists, suppliers of fertilizers and seeding, agro-machinery and other inputs.

#### 1.4.4.2   France

France was once a major lavender oil producer, but with lavandin brought under cultivation; it now is the major producer of lavandin oil of around 1,200 tons per year (IFEAT Lavender report, 2016). The lavender fields of Provence, France, are the most spectacular natural sights, with beautiful purple flowers stretching for miles into the horizon. In 2016, the estimated land cover of about 22,213 hectares of land under 1,496 farm holdings was reported to produce 109 tons of lavender oil and 1,439 tons of lavandin oil (Schmidt & Wanner, 2020).

#### 1.4.4.3   Ukraine, Spain and Japan

In Ukraine, more than 320,000 lavender bushes of various varieties are located on the territory of the Lavender Mountain farm in Perechyn. Brihuega, Spain, is known to have about 30 + dedicated lavender cultivation fields covering an area of over 10,000 hectares. Similarly, even in Japan, about 480 hectares of land is being cultivated under lavender cultivation for the extraction of EO and promoting tourism.

#### 1.4.4.4 India

Lavender cultivation is being promoted and practised in more than 20 districts of Jammu and Kashmir, Uttarakhand and Himachal Pradesh. Under the purple revolution, about 200 hectares of land under the land holding of 1,200 farmers is cultivated, employing about 5,000 associated villagers.

### 1.4.5 MINT OIL – USA AND INDIA

Mint oil is extracted from the whole plant by steam distillation at the flowering stage. Out of 25–30 known mentha species, three main species of *Mentha arvensis*, *Mentha piperita* and *Mentha spicata*, are being commercially cultivated for EO production. The EO content is in the range of 0.38%–2% (Jezler & Mangabeira, 2005). Several farms within the United States, India and China are dedicatedly utilized for peppermint production.

The United States dominates the production and export of peppermint oil, with the mint industry as the largest commercial herb industry. In 2015, 26,500 hectares of land under cultivation were reported to produce about 2,700 tons; in 2017, 24,500 hectares were reported to yield approximately 2,600 tons of peppermint oil. The drop in cultivation land coverage was mainly because of less available production land for peppermint as corn production takes over land and large raw material handling required to produce substantial amounts of EO (about 80 kg oil/ hectare of land) (Gilman et al., 2019).

Similarly, India is the most competitive global supplier of corn-mint EO, accounting for 90% of global crude oil production. In India, in 2009, around 160,000 hectares of land were under corn-mint cultivation, producing around 16,000 tons of mentha oil, with an increase in 2012 to about 34,500 tons and in 2013 to about 50,000 tons (IFEAT Mint Report, 2019). This entire production is mainly cultivated in the states of Haryana, Uttar Pradesh, and Punjab by small and marginal farmers with small land holdings of 0.2–2 hectares of land. Being a short-term crop, mint is mainly cultivated as a third crop or Zaid season between rice and wheat cropping (Kumar et al., 2011). It is harvested twice a year, in April and in August.

An estimated 24,000 tons of oil was produced with the involvement of around 12,750,000 people in 2010. On extrapolating the same in 2012–13, considering the production of 45,000 tons, the number of people estimated to be involved is around 15,000,000. Though it cannot be said that this number of people only depends on mint farming as it is one of the three crops harvested yearly (IFEAT Mint Report, 2019; Kumar et al., 2011).

Though mint cultivation is a short-duration business, it is extremely labour-intensive. Cultivators are looking for other cultivation options, resulting in decreasing available land and less raw material, causing fluctuation in oil prices as high as 150%. This is thus causing a spike in the price of natural mentha oil for consumers who, as a result, are opting for a synthetic menthol replacement (Gilman et al., 2019).

## 1.4.6 Vetiver Oil – Haiti

The EO of vetiver is obtained by distillation of the roots growing as deep as 2–4 metres with a very strong scent. There is also a variety with shallow rooting of about 6–12 inches, which is easier for harvesting. The oil yield may vary from 0.15% to 1% with major constituents such as benzoic acid, vetiverol, furfurol, vetivone and vetivene (Smitha et al., 2014). With a global demand of around 408 tons though indigenous to India, Pakistan, Bangladesh, Sri Lanka and Malaysia, the main producers are Tropical Asia, Africa, Australia, Haiti and Indonesia.

Haiti still retains its status as the site of the world's highest quality vetiver, with being "more balanced and multi-faceted" variety than seen in Indonesia or India. In 2009, when the global demand was about 250 tons, 16 operational refineries were reported to produce 100 tons of oil, approximately 400 drums (249 kg capacity each) with a cultivation land coverage of about 10,000 hectares. It was estimated that 15,000–60,000 farmers rely on vetiver root cultivation as their primary source of income. In Haiti, vetiver is cultivated in small plots by small-scale farmers. Harvesting the root is a critical process; the planter cannot harvest the cultivation plots, creating employment opportunities for more people.

The extracted oil is exported through agents and exporters. Each barrel is valued at $40,000 to $43,000 in the US Market, costing $190 to $200 per kg of oil. The markup of exporters in this price is about $5–$10 (Freeman, 2011). The overall worth of oil exported from Haiti is about $ 16 million to $17 million.

In India, vetiver is mainly cultivated in the states of Rajasthan, Uttar Pradesh, Kerala, Karnataka, Madhya Pradesh and Andhra Pradesh and is wildly gathered in the states of Punjab, Uttar Pradesh and Assam. The estimated production of vetiver oil in India is about 20–25 tons, most of which is being consumed locally, considering the annual demand of approximately 100 tons (Smitha et al., 2014).

## 1.4.7 Frankincense and Myrrh – Somalia

Frankincense and myrrh are both obtained in the form of resins from small trees and shrubs of the Boswellia genus (*Boswellia carterii [Somalia]* and *Boswellia papyrifera [Ethiopia]* giving frankincense) and Commiphora genus (*Commiphora myrrhae* giving myrrh) of the Burseraceae family. Numerous species and varieties of the *Burseraceae* family produce different types of resin. Geographic diversity in soil and climate also has a major impact on the quality of the resin, even within the same species. Frankincense and myrrh are tree resins produced in Ethiopia, Somalia, Somaliland and Puntland countries from East Africa, Yemen and Oman of the southern Arabian Peninsula, and the Sahelian region of Africa.

The resins are used in an unprocessed form for both fragrance and flavour purposes, distilled to yield volatile oils with characteristic and balsamic odours used in perfumery and solvent extracted to get resinoids and absolutes used as fixatives in perfumes. *Boswellia serrata* yields 10% or more, *Boswellia sacra* yield around 9%–10%, and *Boswellia carterii* and *Boswellia frereana* typically yield 5%–6% and 2%, respectively.

The global demand or consumption of the oil or resin is difficult to estimate, considering geographic isolation, and the nomadic nature of much of the collection area, however, is estimated to be around 2,500 tons per year. The reported imports from France, the United Kingdom and the United States alone were at least 400 MT per annum of *B. carterii* or *B. sacra* for application in the fragrance industry.

Considering the lack of clarity on the production and export data, the social impact on the locals is also difficult to be known. Frankincense and Gums Development and Sales Agency estimated that around 10,000 families in northern Somalia primarily depend on gum gathering.

Gathering the gums involves tapping wild-growing trees, after which resin is left to exude from the tree and harden for a few days, then collected every 10–15 days. It is a wildcraft method. The gum collected by the villagers or farmers is stored in an excavation, cave or some kind of camp for stabilization and gathering sufficient quantities to be sold per the classification, gradation and trade. After 12 weeks of hardening, cleaning, sieving, and sorting are carried out at collection centres. This entire manual process involves most of the local populations from the villages and towns. The traders and exporters are associated after this for exporting the product in this form or selling it in local markets for chewing (IFEAT Report Frankincense & Myrrh, 2017).

### 1.4.8 Patchouli – Indonesia

Patchouli is one of the most important EOs used as a base note in perfumery and fine fragrances owing to its unique and complex properties. The leaves are covered with trichomes all over the epidermis, which contains the EO that can be collected by steam distillation of shade-dried leaves. The herb is grown extensively in the tropical climate of Indonesia, Malaysia, Singapore, China and Brazil, preferably under partial shade. The leaves are harvested throughout the year but with peak production in June to July and November to December. The oil content of the shade-dried leaves is around 2.5%–3%, with contents of patchoulol and caryophyllene as major constituents. Indonesia alone accounts for the production of 90% of global patchouli oil demand. Other producing countries include China and India, but with a minor role in global production (Tripathi et al., 2005).

The producing regions in Indonesia are Sulawesi, accounting for 70%; Sumatra, 25%; and Java, 5% of total production. The cultivation is carried out by around 12,000 small-scale farmers with farm holdings of 0.25–1 hectares. The produce of dry leaves is about 4–5 tons per hectare, which helps produce about 25–100 kg of patchouli oil for farmers as per their land under cultivation. The distillation is carried out with the help of around 250 field distillation units in Sulawesi and Java and around 175 in Sumatra. Patchouli oil business provides a livelihood for 50,000 people for cultivation considering 4 people per family, 2,000 people for extraction, considering 5 people per extraction unit and around 300 people for trading. Production reached a level of 1,200–1,400 tons, and the patchouli oil business valuation is about $70–$100 million (IFEAT Report Patchouli, 2015).

Patchouli is a modest proportion of farmers' annual revenue (less than 25% on average). High fluctuations in oil prices and a lack of knowledge of good agricultural processing have demotivated farmers to grow patchouli and encouraged them to undertake other economic activities.

### 1.4.9 BERGAMOT – ITALY

Bergamot oil is a citrus oil extracted from the peels of the bergamot fruit. Bergamot is a hybrid between bitter orange and lemon. There are three varieties of bergamot fruits grown, namely *Feminello*, *Fantastico* and *Castagnaro*. Bergamot is a key ingredient and game changer in more than 50% of all fine fragrances worldwide, and it constitutes the base of cologne water (eau de cologne), making it commercially very important. This fruit is commercially grown primarily for the rind oil, which can be extracted by steam distillation or cold press, yielding about 2.5%–4% oil, mainly constituting limonene, linalyl acetate, linalool, γ-terpinene and β-pinene (Giwa et al., 2018). It is mainly grown in Italy, the Ivory Coast and Brazil.

Calabria in southern Italy accounts for almost 90% of total global production. The cultivation areas initially were just over 3,000 hectares, spreading across a 140 km stretch of land, beginning in Reggio di Calabria, heading south and following the Ionian Coast. Now a further 450 hectares are being planted. Of the total, 1,400 hectares are planted and managed through several cooperatives and growers' associations – the rest by individuals. Around 20% of the total cultivated area is represented by family fields < 2 hectares, 25% have 2–5 hectares, 25% have 5–10 hectares, and 25% of the cultivated land is > 10 hectares. More than 4,500 families are involved in the production cycle of the oil. Approximately 25,000 metric tons of fresh fruit is produced annually for mechanical extraction (Pelatrice extractors). An approximate yield of 1 kg of oil from 200 kg of fruit leads to the current production of 125 metric tons per year, which accounts for more than 3% of Italy's export.

It is also seen that there has been an 11% increase in recent years, and 83% of the growers are ready to expand their production by planting more trees, owing to the ambitious policies helping the entire supply chain, including all stakeholders, from farm to university, processors and users (IFEAT Bergamot oil, 2015).

### 1.4.10 VANILLA IN MADAGASCAR

Vanilla, native to Mexico, was brought to Madagascar during colonization by the Spanish and Portuguese. The absence of pollinating bees makes cultivating vanilla orchids time-consuming as they must be hand pollinated. Once planted, it takes the plant 3 years to bear fruit the first time. Out of 110 variants of vanilla, *Vanilla Planifolia A* is the one that is being cultivated in Madagascar. Being the main exporter, Madagascar's export of vanilla is worth $531 million, which is 62% of the total market in 2018 (Hänke et al., 2019).

The SAVA region in northeastern Madagascar is estimated to produce almost 70%–80% of all global bourbon vanilla on approximately 25,000 hectares of land, with the involvement of 70,000 farmers. This cultivation is by small and marginal farmers practising farming by traditional methods without mechanization. Vanilla is a cash crop that can be immediately sold for large profits. Farmers manually clear plots, plant trees and vanilla vines, weed, and hand pollinate each flower. Besides this, little is known about the farming population, their livelihoods, and the impact of vanilla cultivation on biodiversity (Raxworthy, 2019).

## 1.4.11 CITRONELLA – CHINA AND INDONESIA

Java citronella is an important aromatic grass extensively used in the perfumery, cosmetic and flavouring industries. The key characteristic of the oil is the insect-repellent feature. Citronella oil is classified in trade into Ceylon citronella oil from *Cymbopogon Nardus* and Java citronella oil from *Cymbopogon winterianus*, which is produced and traded in greater volume. The overall oil yield of citronella EO is about 0.8%–1%, consisting of citronellal, geraniol, citronellol, geranyl acetate, neral, geranial, elemol and limonene (Singh & Kumar, 2017). Java citronella oil is mainly produced in China, about 800 and 1,500 metric tons, and in Indonesia, about 250 and 500 metric tons, to meet the annual global requirement of around 1,800 metric tons.

In China, cultivation of citronella is carried out in remote and underdeveloped regions of southern and far western regions of Yunnan Province, particularly in Luchun County, Honghe Prefecture, Mojiang, Simao Prefecture and Yingjiang County, Dehong Prefecture. Farmers undertake cultivation followed by field distillation as an important cash income source. The production of 800–1,500 MT of oil is produced by a total cultivation area of about 6,600 hectares, with 20–30 thousand people engaged in the citronella business.

It is mainly grown in Java and West and North Sumatra in Indonesia. In Indonesia, the average annual production of 400 metric tons is produced from 2,000 hectares of land, generating a livelihood for 5,000 people (1,000 farming families, extractors and agents). These farmers are small-scale farmers with an average landholding of 2 hectares. The average harvest yields 10 metric tons of citronella grass per hectare per annum, which is distilled to get about 100 kg of oil (IFEAT Citronella Report, 2014).

## 1.4.12 GERANIUM – CHINA AND EGYPT

Geranium oil is extracted from the fresh leaves and stalks of the plant by using steam distillation. Out of 270 known species of *Pelargonium*, Rose geranium *(Pelargonium rosé)* is a hybrid species developed from crossing *P. capitatum* with *P. radens* for commercial oil extraction. Fresh leaves yield about 0.15%–2% of greenish olive bourbon oil, which can be blended with lavender, patchouli, clove, rose, orange blossom, sandalwood, jasmine, juniper, bergamot and other citrus oils owing to its rosy sweet minty smell. It mainly consists of citronella, geraniol, linalool, iso-menthone, menthone, phellandrene, sabinene and limonene as major constituents (Department of agriculture, 2012). With a global demand of 350–400 tons, China with 80–100 tons and Egypt with 200–230 tons meet the maximum production. Other countries that produce smaller quantities to meet the remaining 20% or less are India, about 25–35 tons/year; Madagascar < 10 tons/year; South Africa, about 5–10 tons/year; Reunion Island, about 2–6 tons/year; Kenya, about < 1 ton/year; Morocco, about < 0.5 tons/year; and Congo, about < 0.5 tons/year.

The geranium oil industry provides a livelihood for about 25,000–30,000 people – including intermediaries, transport workers, factory workers, exporting companies and 5,000–7,000 farming families in China. Similarly, about 30,000–35,000 people, including 8,000 farming families and the whole supply chain, benefitted in Egypt. On extrapolating, one can assume that the economic benefits of the geranium oil industry are being shared by about 100,000–150,000 people globally (IFEAT Geranium Report, 2015).

Though in the previous section it is mentioned that lack of marketing network and price fluctuation is not encouraging farmers to pursue the cultivation of aromatic plants, in this section, it can be seen that across the globe, countries have developed identities for the export of particular EOs. It can also be seen that aromatic crop cultivation can be taken up in a good-enough land spread, influencing the lives of thousands of farming families. Aromatic cultivation is largely taken up by small and marginal farmers (poor and underprivileged), representing a key income generator for those farmers. It could thus help reduce poverty and increase investment in health and education services that are vital for overall economic development and social climbing. The only requirement seen now is to have an organized network for providing quality plantation material, educating the cultivators, ensuring localized extraction units, having centralized processing and testing facilities and having fair trade practices ensuring sustainable income for the farmers and consistent supply for the industry.

## 1.5  CONCLUSION

The complexity of the plant-based aroma and EOs, considering the chemical composition, is well recognized. Since a selective compound in the oil plays a critical role in the respective application, extraction of the same in primitive form is crucial. The secondary metabolite content of plant biomass is very small; hence, significant attention is required towards cultivation, harvesting and post-harvesting practices like drying and extraction methodology. Considering the trade data and the industrial demand for these natural volatiles and EOs, huge cultivation grounds are required with well-researched processing protocols.

Having said so, aromatic plant cultivation is a labour-intensive industry and is seen to provide significant employment. It is seen that a proper network and collective cultivation of the aromatic crop in countries like Paraguay, Turkey, Bulgaria, Iran, Ukraine, France, India, the United States, Haiti, Italy, Madagascar, Egypt and China have managed to impact a significant population and development of a sustainable global market for the produce.

## REFERENCES

Bhaskara Rao, T. S. S., & Murugan, S. (2021). Solar drying of medicinal herbs: A review. *Solar Energy*, *223*(June), 415–436. 10.1016/j.solener.2021.05.065

Bleimann, K. (2019). Rose oil: history and socio-economic impact. *Perfumer & Flavorist*, *44*, 51–54.

Boukhris, M., Ben Nasri-Ayachi, M., Mezghani, I., Bouaziz, M., Boukhris, M., & Sayadi, S. (2013). Trichomes morphology, structure and essential oils of Pelargonium graveolens L'Hér. (Geraniaceae). *Industrial Crops and Products*, *50*, 604–610. 10.1016/j.indcrop.2013.08.029

Calín-Sánchez, Á., Lipan, L., Cano-Lamadrid, M., Kharaghani, A., Masztalerz, K., Carbonell-Barrachina, Á. A., & Figiel, A. (2020). Comparison of traditional and novel drying techniques and its effect on quality of fruits, vegetables and aromatic herbs. *Foods*, *9*(9). 10.3390/foods9091261

Chalova, V. I., Manolov, I. G., & Manolova, V. S. (2017). Challenges for commercial organic production of oil-bearing rose in Bulgaria. *Biological Agriculture and Horticulture*, *33*(3), 183–194. 10.1080/01448765.2017.1315613

DAFF. (2009). *Lavender production* (Vol. 353, Issue 0).

Department of agriculture, F. and F. (2012). Rose geranium production. In *Directorate: Plant Production*. www.daff.gov.za/publications

Deshingkar, P., & Akter, S. (2009). Migration and human development in India. *Human Development Research Paper (HDRP) Series*, *13*(19193). http://mpra.ub.uni-muenchen.de/19193/

Dhifi, W., Bellili, S., Jazi, S., Bahloul, N., & Mnif, W. (2016). Essential oils' chemical characterization and investigation of some biological activities: A critical review. *Medicines*, *3*(4), 25. 10.3390/medicines3040025

Figueiredo, A. C. (2017). Biological properties of essential oils and volatiles: Sources of variability. *Natural Volatiles and Essential Oils*, *4*(4), 1–13.

Firn, R. (2010). Nature's chemicals: Natural products that shape our world. In D. Chen (Ed.), *Oxford Biology* (Vol. 4). Oxford University Press. 10.14237/ebl.4.2013.18

Freeman, S. (2011). Vetiver in Southwest Haiti. In *Columbia University*. http://haiti.ciesin.columbia.edu/haiti_files/documents/Freeman_UNEP_Vetiver_Report_2011_0.pdf

Gade, D. W. (1979). Petitgrain from Citrus aurantium: Essential Oil of Paraguay. *Economic Botany*, *33*(1), 63–71. http://www.jstor.org/stable/4254012

Gilman, F., Keyari, C., Kropp, R., Speight, L., & McLear, C. (2019). Bio-based butanol as a solvent for essential oil extractions. *Perfumer & Flavorist*, *44*(March), 27–35.

Giray, F. H., & Omerci Kart, M. C. (2012). Economics of Rose Demascena in Isparta, Turkey. *Bulgerian Journal of Agriculture Sciences*, *18*(5), 658–667.

Giwa, S. O., Muhammad, M., & Giwa, A. (2018). Utilizing orange peels for essential oil production. *ARPN Journal of Engineering and Applied Sciences*, *13*(1), 17–27.

Hänke, H., Bührmann, A. D., Franke, Y., & Marggraf, R. (2019). *Socio-economic, land use and value chain perspectives on vanilla farming in the SAVA Region ( north-eastern Madagascar): The Diversity Turn Baseline Study (DTBS)* (Issue July). 10.13140/RG.2.2.22059.80163

Huchelmann, A., Boutry, M., & Hachez, C. (2017). Plant glandular trichomes: Natural cell factories of high biotechnological interest. *Plant Physiology*, *175*(1), 6–22. 10.1104/pp.17.00727

IFEAT Bergamot oil. (2015). *Socio Economic impact study of Naturals - Bergamot oil.*

IFEAT Citronella Report. (2014). *IFEAT Socio Economic impact study - Citronella* (Issue May).

IFEAT Geranium Report. (2015). *An overview of important Essential oils and other Natural Products - Geranium.*

IFEAT Jasmine report. (2015). *An overview of important Essential oils and other Natural Products - Jasmine* (Issue November).

IFEAT Lavender report. (2016). *An overview of important Essential oils and other Natural Products - Lavender* (Issue October).

IFEAT Mint Report. (2019). *IFEAT Socio Economic impact study - Mint.*

IFEAT Report Frankincense & Myrrh. (2017). *An overview of important Essential oils and other Natural Products - Frankincense and Myrrh.*

IFEAT Report Patchouli. (2015). *Socio Economic Committee report on Patchouli - Pogostemon cablin.*

IFEAT Rose report. (2019). *Socio Economic impact study of Naturals - Rose.*

Jezler, C. N., & Mangabeira, P. A. O. (2005). The yield and essential oil content of mint (Mentha spp.) in Northern Ostrobothnia. In *Acta Horticulturae* (Issue 1). 10.1080/00103624.2010.504798

Jin, W., Mujumdar, A. S., Zhang, M., & Shi, W. (2018). Novel drying techniques for spices and herbs: A review. *Food Engineering Reviews*, *10*, 34–45. 10.1007/s12393-017-9165-7

Joy, P. P. (2007). Aromatic Plants. In K. V. Peter (Ed.), *Aromatic Plants* (1st ed., Issue July). New India Publishing Agency. 10.1007/978-94-009-7642-9

Khan, M. I. H., Wellard, R. M., Nagy, S. A., Joardder, M. U. H., & Karim, M. A. (2017). Experimental investigation of bound and free water transport process during drying of hygroscopic food material. *International Journal of Thermal Sciences*, *117*, 266–273. 10.1016/j.ijthermalsci.2017.04.006

Kovacheva, N., Rusanov, K., & Atanassov, I. (2010). Industrial cultivation of oil bearing rose and rose oil production in Bulgaria during 21ST century, directions and challenges. *Biotechnology and Biotechnological Equipment*, *24*(2), 1793–1798. 10.2478/V10133-010-0032-4

Kumar, S., Suresh, R., Singh, V., & Singh, a. K. (2011). Economic analysis of menthol mint cultivation in Uttar Pradesh: A case study of Barabanki district. *Agricultural Economics Research*, *24*(December), 345–350.

Lubbe, A., & Verpoorte, R. (2011). Cultivation of medicinal and aromatic plants for specialty industrial materials. *Industrial Crops and Products*, *34*(1), 785–801. 10.1016/j.indcrop.2011.01.019

Majumder, P., Sinha, A., Gupta, R., & Sablani, S. S. (2021). Drying of selected major spices: Characteristics and influencing parameters, drying technologies, quality retention and energy saving, and mathematical models. *Food and Bioprocess Technology*, *14*, 1028–1054.

Orphanides, A., Goulas, V., & Gekas, V. (2016). Drying technologies: Vehicle to high-quality herbs. *Food Engineering Reviews*, *8*(2), 164–180. 10.1007/s12393-015-9128-9

Pise, V., Shirkole, S., & Thorat, B. N. (2022). Visualization of oil cells and preservation during drying of Betel Leaf (Piper betle) using hot-stage microscopy. *Drying Technology*. 10.1080/07373937.2022.2048848

Prothon, F., Ahrne, L., & Sjoholm, I. (2003). Mechanisms and prevention of plant tissue collapse during dehdration - A Critical review. *Critical Reviews in Food Sci*, *43*(4), 447–479.

Qiu, L., Zhang, M., Mujumdar, A. S., & Liu, Y. (2020). Recent developments in key processing techniques for oriental spices / herbs and condiments: A review. *Food Reviews International*, 1–21. 10.1080/87559129.2020.1839492

Radoj, M., Pavkov, I., Kovacevic, D. B., Putnik, P., Wiktor, A., Stamenkovic, Z., Kešelj, K., & Gere, A. (2021). Effect of selected drying methods and emerging drying intensification technologies on the quality of dried fruit: A review. *Process*, *9*(132), 21.

Rahman, M. M., Kumar, C., Joardder, M. U. H., & Karim, M. A. (2018). A micro-level transport model for plant-based food materials during drying. *Chemical Engineering Science*, *187*, 1–15. 10.1016/j.ces.2018.04.060

Raul, A. (2020, January). Characteristics & socio-economic impact of petitgrain oil Paraguay. *Perfumer & Flavorist*, *45*, 28–30.

Raxworthy, T. (2019). *Vanilla Bean Farming in Madagascar: An Economic and Social Report of Policy and Development Vanilla Bean Farming in Madagascar* (Issue May).

Rehman, R., & Asif Hanif, M. (2016). Biosynthetic factories of essential oils: The aromatic plants. *Natural Products Chemistry & Research*, *04*(04). 10.4172/2329-6836.1000227

Sanganeria, S. (2014). Vibrant India- Opportunities For The Flavour And Fragrance Industry. In *Ultra International Limited* (Vol. 30, Issue 7). %5C%5CRobsrv-05%5Creference manager%5CArticles%5C10532.pdf

Saripalle, M. (2016). Jasmine cultivation in Tamil Nadu: Market structure and pricing. *World Development Perspectives*, *1*, 12–14. 10.1016/j.wdp.2016.05.004

Schmidt, E., & Wanner, J. (2020). Adulteration of essential oils. *Handbook of Essential Oils*, *July*, 543–580. 10.1201/9781351246460-20

Sharifi-Rad, J., Sureda, A., Tenore, G. C., Daglia, M., Sharifi-Rad, M., Valussi, M., Tundis, R., Sharifi-Rad, M., Loizzo, M. R., Oluwaseun Ademiluyi, A., Sharifi-Rad, R., Ayatollahi, S. A., & Iriti, M. (2017). Biological activities of essential oils: From plant chemoecology to traditional healing systems. In *Molecules* (Vol. 22, Issue 1). 10.3390/molecules22010070

Simoes, A., & Hidalgo, C. (2021). *Observatory of Economic Complexity (OEC) - The Economic Complexity Observatory: An Analytical Tool for Understanding the Dynamics of Economic Development*. Workshops at the Twenty-Fifth AAAI Conference on Artificial Intelligence.

Singh, A., & Kumar, A. (2017). Cultivation of Citronella (Cymbopogon winterianus) and evaluation of its essential oil, yield and chemical composition in Kannauj region. *International Journal of Biotechnology and Biochemistry*, *13*(2), 139–146. https://www.ripublication.com/ijbb17/ijbbv13n2_04.pdf

Smitha, G. A., Varghese, T. S., & Manivel, P. (2014). Cultivation of vetiver. In *Indian Council of Agricultural Research*.

Thamkaew, G., Sjöholm, I., & Galindo, F. G. (2021). A review of drying methods for improving the quality of dried herbs. *Critical Reviews in Food Science and Nutrition*, *61*(11), 1763–1786. 10.1080/10408398.2020.1765309

Tisserand, R., & Young, R. (2014). Essential oil composition. In *Essential Oil Safety* (2nd ed.). © 2014 Robert Tisserand and Rodney Young. 10.1016/b978-0-443-06241-4.00002-3

Tripathi, Y. C., Kaushik, P. K., & Pandey, B. K. (2005). Patchouli (Pogostemon cablin): A promising medicinal and aromatic crop for North East India. In *Recent Progress in Medicinal Plants: Plant Bioactives in Traditional Medicines* (1st ed., Issue January, pp. 289–310). Studium Press. 10.13140/RG.2.1.2914.7284

Uthpala, T. G. G., Navaratne, S. B., & Thibbotuwawa, A. (2020). Review on low-temperature heat pump drying applications in food industry: Cooling with dehumidification drying method. *Journal of Food Process Engineering*, May, 13. 10.1111/jfpe.13502

Zhang, Q. W., Lin, L. G., & Ye, W. C. (2018). Techniques for extraction and isolation of natural products: a comprehensive review. *Chinese Medicine*, 1–26. 10.1186/s13020-018-0177-x

# 2 Cellular Structures of Aromatic Plant Materials

## 2.1 INTRODUCTION

Internal structures of a plant can be defined based on levels of organization in the plant anatomy. Cells are the basic units organized in tissues and, in turn, organized into organs. Cells may appear round, elongated, rectangular, polyhedral, kidney-shaped, globular, star-shaped or drum-shaped. Based on the functionality of each tissue or organ, differences in internal structure and the organ's adaptation to the diverse environment are prominently seen. Any resulting change in a chemical composition, shape and structure, organization, and final form is a new challenge for processing – drying and extraction. The histology of the plants and the functioning of the cells and tissues play a significant role, and an understanding of the structure could support the establishment of drying and extraction protocols.

The internal structures can be simplified and described as vacuoles, cytoplasm, cell walls and inter-cellular spaces. *Vacuoles* are an aqueous solution of sugar, organic acids and salts. *Cytoplasms* are gel matrices or complex fluids containing reserve starch, lipids, proteins and cell organelles. Tonoplast and plasmalemma are the protein-lipid membranes that bind and regulate environmental contact. *Cell walls* are the non-static organelles crucial for adjusting to cell growth, metabolism, attachment, shape and stress. These are mainly composed of *hemicellulosic* interlocking components with *micro-cellulosic fibrils* (about 50%–65% weight on a dry basis) embedded in *pectin* substance (about 30%) and *extensin* cross-linking to give the fixed shape. These cells are glued together with a thin layer of the middle lamella, *pectinous* in nature and resulting in the texture of the tissue.

Similarly, the *plasmodesmata* and *cytoplasmic* connection link are the linking agents creating the *intercellular continuum*. This acts as the transport channel for water, small molecules and ions (Prothon et al., 2003). These biomolecules in complex biostructures are difficult to quantify or describe completely. However, they play a critical role in optimizing the dehydration process and retaining structural integrity. The structures, functions and properties of tissues highly affect the mass transfer phenomenon during their life functions (Le Maguer et al., 2003) and are significant for us for post-harvest processing. A basic understanding and anatomical description of these tissues' internal structures, functioning and chemical compositions is crucial for an effective processing method and *dehydration* while retaining the structural integrity of the matrix and *extraction* and separating the secondary metabolites. This chapter briefly explains different cells and their functions in aromatic plants.

DOI: 10.1201/9781003315384-2

## 2.2  PLANT TISSUES

The plants' organs are organized considering three tissue layers: the epidermis, the ground tissue and the vascular system. A group of cells from a common origin and having common functions are termed *tissues* and can be categorized as *meristematic* and *permanent tissues*. *Meristem* is the specialized region of active cell division occurring at the tips of roots and shoots and is largely maintained as different tissues throughout plant development. *Primary meristems* (*apical* and *intercalary*) are early-appearing tissues contributing to body formation. *Secondary meristems* are *lateral meristems* in the mature region producing secondary tissues. After the required divisions, the newly formed cells become structurally and functionally specialized and are termed *permanent* or *matured cells* constituting permanent tissues.

These permanent tissues, having all similar cells and the same function in structure, are simple tissues – parenchyma, collenchyma and sclerenchyma. Similarly, permanent tissues with structures with more than one cell type and working together as a unit are called *complex tissues,* xylem and phloem. The three layers formed for any plant organs through these tissues include the epidermis layer, which is responsible for plant protection and gas exchange; subepidermal tissues, generating most photosynthetically active tissues; and the vascular system, required for long-distance transport of water and metabolites (Hülskamp & Schnittger, 2012).

The dermal tissue is the outermost cell layer that mediates the interactions of a plant with its environment. In most plants, the dermal tissue consists of a single-cell layer. Several specialized cell types execute the various functions of the epidermis. In the aerial parts of a plant, the majority of the epidermal cells are small and compact, with a cuticle consisting of cutin and wax. This cuticle effectively protects the plant from water loss and functions as a barrier against pathogens. The exchange of water vapours and gases is regulated by stomata, small-gated pores formed by the guard cells and their subsidiary cells. Turgor changes in guard cells control stomatal opening and closing and thereby regulate the optimal uptake of carbon dioxide for photosynthesis and the overall water economy of the plant.

The ground tissue has a variety of functions, including photosynthesis, storage, reproduction and mechanical support. Three specific tissue types are classified according to their cell wall structure and thickness and their role in mechanical support within the ground tissue: parenchyma, collenchyma and sclerenchyma cells.

The vascular tissue is a water- and food-conducting tissue that forms a continuous system throughout the entire plant body. It is also important for distributing plant hormones (e.g., auxin) and other signalling molecules. The vascular system consists of two different tissues: the xylem and the phloem.

### 2.2.1  SIMPLE AND COMPLEX TISSUES

*Parenchyma* contains the most primitive, fundamental and ground tissues performing all major functions. These are the most abundantly found tissues in the plant, with thin cell walls made of cellulose and hemicellulose, intercellular space and vacuolated active protoplast with various shapes. These cells and cell walls consist of primary pit fields interconnected by plasmodesma interconnections. *Chlorenchyma* is a sub-class of parenchyma in higher plants, found in the mesophyll of leaves, pericarp of unripe fruits, and cortex of young stems and branches. Similarly, *Arenchyma* is abundantly found in hydrophytes, having large intercellular spaces giving buoyancy and aiding respiration and gas exchange. *Storage parenchyma* is abundant in storage organs like fruits, seeds, tubers, and so on. *Idioblastic parenchyma* stores tannins, oil, and inorganic crystals in succulent xerophytes using cells secreting hydrophilic mucilaginous substances that hold large amounts of water. As epidermal cells, parenchyma gives protection. Turgid parenchyma gives mechanical support to herbs and hydrophytes.

*Collenchyma* are the living mechanical tissues in stems and leaves, aiding the function of protection. Compared to parenchyma, these are arranged irregularly without intercellular spaces owing to unevenly thickened cell walls with excessive deposition of cellulose and pectin. They are also characterized by a high pectin content, about 60% water in cell walls, and vacuolated protoplast. The characteristic collenchyma position is hypodermal, as a continuous or discontinuous ring, and as *angular, lacunar or lamellar* type, providing mechanical strength, elasticity and flexibility.

*Sclerenchyma* is a widely distributed, important mechanical tissue consisting of dead and empty cells with reduced lumen with highly thickened and lignified walls. *Sclerenchyma* is of mainly two types, *fibres*, elongated cells with tapered ends and *sclerides,* short cells. However, both these types are characterized by thickened, lignified and hard cell walls (Figure 2.1).

*Xylem* are the complex tissues that conduct water and minerals. Xylems are mostly dead tissues except for the parenchyma. The water and dissolved salts conducting elements are *tracheid* and *vessels*. The *tracheids* are elongated cells with tapered ends, lignified walls, and narrow lumen. They have imperforated end walls and bordered pits on lateral walls. Vessel members are elongated but cylindrical

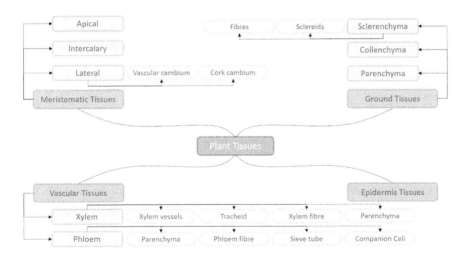

**FIGURE 2.1** Plant tissue classification.

cells with lignified cell walls and wide lumen. In the case of vessel members, end walls are oblique and perforated. These tracheids and vessels both show several types of wall thickening, such as annular, spiral, scalariform, reticulate and pitted, increasing the complexity of the structure's composition. *Xylem fibres* have similar composition of tracheid and vessels but are elongated spindle-shaped dead cells with sharp and tapered ends. *Xylem parenchyma* are living cells with thin walls filled with vacuolated and nucleated cytoplasm. These are flexible cells with non-lignified cell walls in rectangular shapes. These are further classified as *primary xylem* (consisting of only axial system) and *secondary xylem* (consisting of axial and ray systems). These parenchyma cells sometimes outgrow into *tyloses,* which tend to block water movement. This is typically seen when sapwood is converted to heartwood or under the stress of pathogenic fungi to check the spread.

*Phloem* are living cells that conduct food materials. These tissues consist of *sieve elements (cells & tubes), companion cells* and *phloem parenchyma,* which are living cells, and *phloem fibres,* which are non-living cells. *Sieve elements* are vacuolated protoplast, highly specialized chief conducting cells. The presence of a sieve area in the cell walls is the main characteristic feature consisting of numerous minute pores. *Companion cells* are elongated rectangular parenchyma cells associated with sieve elements. *Phloem fibres* are the only dead cells in the phloem tissues that give mechanical strength to the vascular bundle. Phloem parenchyma is rectangular-shaped walled cells with vacuolated protoplasm consisting of cytoplasm with starch, fats, tannins, and resins.

### 2.2.2 Special Tissues

*Special tissues*, also called secretory or glandular tissues, are located at different parts and are widely distributed in different plant genres. They are very specific to the occurrence type of secretion and place of storage. A few special tissues can be listed as *digestive glands* (in the case of insectivorous plants), *nectary glands* (in the case of floral and extrafloral nectaries), *osmophores* (in the case of aromatic plants), *secretory cavities* (for storage and release of secretions – *lysigenous cavities* and *schizogenous cavities*), *hydathodes* (water secreting structures), and *laticiferous tissues* (latex secreting tissues). In aromatic plants, the main tissues that need to be understood are *osmophores and secretory cavities* and, to some extent, *laticiferous tissues.*

As seen in Chapter 1, essential oils are important organic molecules with natural essence and are biosynthesized by plants. The function of secretion of these secondary metabolites by aromatic plants is seen in specialized cells such as *osmophores, glandular trichomes* and *ducts & cavities.* These organic compounds are synthesized, transported for storage and transported for release into the atmosphere. Aromatic plant cells are very diverse in morphology and range from highly specialized trichomes, ducts, cavities, secreting trichomes, conical-papillate cells, and other essential oil secretory tissues like osmophores and secretory cells (Rehman & Asif Hanif, 2016). To preserve/selectively extract the secreted metabolites in/from the natural matrix during processing, it is vital to understand the cell and tissue structures and functioning in aromatic plants.

## 2.3   HISTO-ARCHITECTURE OF AROMATIC PLANT PARTS

The chemical structures of the primary cell walls of the different parts of aromatic plants and their progenitors differ for different plant species. They vary in the complex glycans that interlace and cross-link the cellulose microfibrils to form a robust framework, in the nature of the gel matrix surrounding this framework, and in the types of aromatic substances and structural proteins that covalently cross-link the primary and secondary walls and lock cells into shape (Carpita, 1996). The chemistry of structural elements like polysaccharides, aromatic substances, and proteins of different plant parts is unique and results in synthesizing and assembling dynamic and functional cell walls (Megías et al., 2022).

### 2.3.1   STRUCTURES OF ROOTS

The primary function of the rooting system is meant for anchoring and absorbing. Root growth is seen by cell proliferation and elongation produced by the apical root meristem. Water and organic molecules transfer through the roots through the vascular system several millimetres from the meristem tissues. Roots grow under *the primary growth mechanism, increasing the length and under the secondary growth mechanism* growing in diameter. This structure is more or less quite similar structure along the root extension. Primary roots are relatively less complex structures forming a layer of epidermal and hypodermis cells under the epidermis, followed by the parenchyma cells. The vascular bundles are seen in the inner bundle of the roots in the form of di-, tri-, tetra- or poly arch organization of separate and alternate rows of xylem and phloem. The secondary roots grow with the *pro-cambium meristem* between the xylem and phloem, becoming the continuous *vascular cambium* along the axis. These are followed by the formation of secondary xylems and phloem, pushing the vascular cambium towards the surface of the roots and increasing the thickness of the roots.

### 2.3.2   STRUCTURES OF STEM AND BARK

Stems are the aerial part of plants, mainly functioning for supporting the organs and transporting or conducting substances, photosynthesis and storage. Similar to the roots, stem growth begins through *caulinar meristem*. Stems are more complex structures with tidily arranged nodes, internodes and axillary buds. Similar to roots again but more complex, the growth is seen as the *primary mechanism*, leading to all stem tissues, leaf primordia and axillary buds, and the *secondary mechanism* is an increase in diameter. The shoot apical meristem, intercalary meristem, secondary meristems; vascular cambium and cork cambium; and lateral meristem are mainly responsible for the primary and secondary growth.

The primary growth of stems usually involves the *epidermis* and *cortex*. The epidermis is a single cell thick, covering the stem and showing *cutin* and *suberin* macro-molecules in the free cell wall. The cortex is a thicker layer of parenchyma cells that can perform photosynthesis or storage. The cortex has hypodermis as the closest cell layer to the epidermis, followed by the vascular cambium activity and the elongation of the vascular cell. It contains the sclerenchyma and collenchyma as the support tissues and occasionally sclereids, glandular cells, and laticifers. The vascular bundle are collateral bundles with primary phloem at the inner core, and primary xylem localized externally. These are again scattered in parenchyma tissues.

The organizations of stem and root vascular bundles are different but are connected at the transition region. The xylem and phloem alternate in the root, and the meta-phloem is inner to the proto-phloem. In the shoot, the phloem is outer to the xylem, and the proto-xylem is inner to the meta-xylem. The phloem and xylem change positions, and the xylem shows a twist.

### 2.3.3 STRUCTURES OF WOOD

The secondary growth of stems is due to the vascular cambium meristem and cork meristem, known to increase the thickness of the stem. The vascular cambium meristem is differentiated from both procambium (or fascicular cambium) and interfascicular parenchyma during the transition from primary to secondary. The vascular cambium develops into a cylinder that, by proliferation and differentiation, gives secondary phloem outward and a secondary xylem inward. Hence, the previous primary vascular tissues, primary xylem and primary phloem are pushed away from each other and remain as small groups of cells at the surfaces of the secondary vascular tissues. The vascular cambium activity imprints signs of a growing ring every year, which are inner as the meristem moves away from the central axis of the stem. This mechanism produces growth in the thickness of the stem. The older differentiated cells from the vascular cambium are the innermost cells of the stem, whereas the more recent differentiated cells are those closer to the meristem. These are typically heartwood and sapwood. In woody stems, another meristem known as phellogen or cork cambium produces the periderm or bark, which replaces the epidermis.

The stem represents the structures as periderm, bark, and vascular tissues – secondary phloem, vascular cambium, secondary xylem and pith or medulla.

*Periderm* is the outer part of the stem and functions as a protective structure. The other component of the periderm is the phelloderm, a layer of tissue inner to the cork cambium. Mostly the cork cambium appears after the vascular cambium, also called a cork or phellem.

*Bark*, accounting for 9%–15% of the stem volume, is the protective layer of the stem, inclusive of the phloem, ritidome (successive periderms) and periderm. It results from two activities of the vascular cambium and cork cambium.

*Vascular tissues* include the secondary phloem produced by the vascular cambium towards the outer surface. The older secondary phloem degenerated and became a part of the bark. In contrast, the new phloems consisting of parenchyma cells, sieve tubes and companion cells are close to the vascular cambium. The vascular cambium is the lateral meristem parallel to the surface of the shoot. It is responsible for the formation of secondary phloems on the outside and xylems on the inside. Secondary xylems are the wood-forming tissue composed of tracheae, tracheids, sclerenchyma fibres and parenchyma cells. The inner layers of these secondary xylems are dead and non-functional as the parenchyma cells die and the chemical composition of the cell wall changes. The storage space of these cells is infiltrated with oils, tannins and resins.

### 2.3.4 STRUCTURES OF LEAVES

Leaves are mainly divided into *blades* and *petioles*. The majority of stomata and photosynthetic parenchyma are found in the blade. These blades are a complex network of *veins* – *primary veins* and *secondary veins*. They combine with intramarginal veins and are divided into tertiary and quaternary veins (*areole*). This complex system of veins is known as reticulate venation and feature dicot leaves forming the transport network for water and gases. The transport occurs through these vein networks finally to a free veinlet within an areole that represents the final level of foliar venation and may be composed of a few tracheids or enlarged cells with thick walls of plant tissue. There are two surfaces in the blade: adaxial (upper), having the vascular bundles arranged with xylems, and abaxial (lower), having vascular bundles arranged with phloem. These are covered under the layer of epidermal cells. The margin of the leaf or leaf contour may show a wide variety of forms. Petiole is more or less long and cylindrical. It connects the blade through the mid-rib with the stem protecting mainly the transport network. Axillary buds, found in the angle between the petiole and the stem, will develop into lateral branches.

### 2.3.5 STRUCTURES OF FLOWERS

The structure of flowers is very abundant and diverse; however, on a more straight-forward plan, it can be understood considering the flower structure of angiosperm flowers. The development begins with the meristem activity of a flower meristem or inflorescence meristem. A typical flower develops on four components of petals, sepals, stamens, and carpels (pistil). The histological organization of petals and sepals is similar to a leaf. Petals show a wide morphological and chromatic variety. Sepals are the main protective structures of the flower. A typical stamen consists of a filament with the anther at the free end. Carpels are highly modified leaf structures combining to form pistils.

Flowers are the reproductive organ of most plants. Seed plants, gymnosperms (inflorescences producing) and angiosperms (flowers after fertilization from seeds enclosed in a fruit). The tissue structure of the flower can be generalized to be whorls of modified leaves in the forms of calyx, corolla, androecium, and gynoecium.

### 2.3.6 STRUCTURES OF SEEDS AND FRUIT

The seed develops from the ovule of the flower ovary. The development begins after fertilization of the egg cell by the microspore of the grain pollen. A typical seed consists of an embryo, endosperms (nurturing tissues wrapping the embryo) and coats. Endosperm cells are storage cells meant for starch or proteins. Seed coats are developed for the tissues surrounding the egg cells. It mainly consists of tegmen and testa firmly attached.

Like the seeds, the ovary walls of the flower are later transformed into fruits. The histological organization of the ovary was similar to a leaf with an inner and an outer epidermal layer and parenchyma with vascular bundles in between. The development of these layers becomes the pericarp, which is actually the fruit without the seed. The pericarp consists of the exocarp, mesocarp and endocarp. The exocarp is the outer layer of the fruit, and the endocarp covers the seed. The tissue between these two layers is the mesocarp, made up of storing parenchyma or sclerenchyma. There are plant species where other parts of the flower, besides the ovary, contribute to form the fruit.

### 2.3.7 STRUCTURES OF THE RHIZOME

The rhizome consists of the epidermis, cortex, and stele. The epidermis has leaf scars, and the outer walls of the epidermis are suberized later in development. The cortical cells divide periclinal toward the periphery, forming a multi-layered hypodermis consisting of rectangular thin-walled cells. The cortex and stele are separated by a single layer of compactly arranged cells, the endodermis, which does not exhibit meristematic activity. The pericycle comprises small and compactly arranged cells situated internally to and contiguous with the endodermis. The pericycle appears to induce growth via tangential and radial division. Vascular bundles are also seen from the pericycle and appear collaterally, forming a ring in the pericycle. The newly generated cells and vascular bundles enlarge and push toward the centre, resulting in continuous primary thickening of the stele while the cortex width remains consistent. Later, many scattered vascular bundles are distributed in the cortex and pith, with a higher density in the pith than in the cortex. Vascular bundles are surrounded by sclerenchymatous cells forming the bundle sheath. Xylem vessels are highly lignified with scalariform perforation plates, and they degrade with the development of the surrounding parenchyma cells containing organelles, including the endoplasmic reticulum and mitochondria (H. Liu & Specht, n.d.) (Figure 2.2 and 2.3).

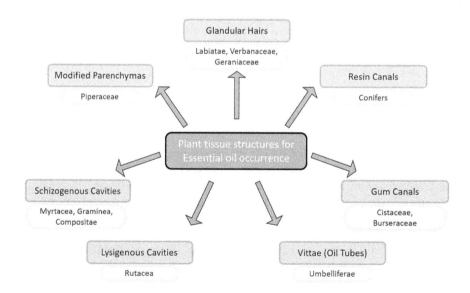

**FIGURE 2.2**  Plant tissues for essential oil occurrence in the different plant families.

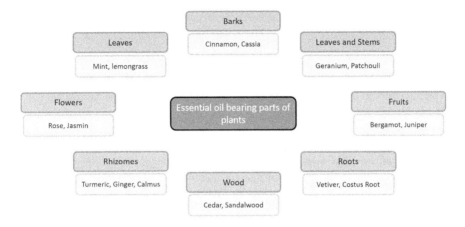

**FIGURE 2.3**   Essential oil-rich plant parts for some aromatic plants.

## 2.4   SPECIAL TISSUES IN AROMATIC PLANTS

The production of essential oils is usually associated with *specialized cell structures*. The formation of these organic molecules in the cells, transfer to the storage tissues, and then release to the atmosphere through the ducts is typically associated with *osmophores, conical papillate cells, glandular trichomes, ducts, cavities* and occasionally *non-organized cells*. Various species and plants of commercial interest have been investigated by research groups of biological and pharmacological departments on secretory structures and factors influencing their development. Secretion involves the discharge of substances to the exterior (*exotropic secretion*) or into special intercellular cavities (*endotropic secretion*). The secreted material may contain various salts, latex, waxes, fats, flavonoids, sugars, gums, mucilages, essential oils and resins. The aroma-contributing compounds of essential oils (with or without resins and gums) are most commonly found in unique secretory structures either transferred from trichomes through the ducts and cavities. These secretory structures vary with the family or species. State-of-the-art instruments like light microscopy (LM), scanning electron microscopy (SEM), and transmission electron microscopy (TEM) have already been used to report the structures and characteristics of these cells.

### 2.4.1   Osmophores

*Osmophores* were defined in 1962 for floral tissues emitting scents. The term *osmophore* means odour-bearing and is also termed a *floral fragrance gland*. These are specialized clusters of cells in flowers that are spread on sepals and petals to attract insect pollinators (Antoń et al., 2012; Curry et al., 1991). These consist of multi-layered glandular epithelium with homogeneous layers of cells. Morphological characteristics of the *Mirabilis Jalapa* flower surface (Effmert et al., 2005), of Stanhopea and Sievekingia (Antoń et al., 2012), and Galanthus nivalis L (Weryszko-Chmielewska & Chwil, 2012) have been studied and well reported.

These cells contain dense cytoplasm, enormous deposits of starch, or other storage compounds within the mesophyll. These deposits are usually missing in epidermis cells. This generates a distinction between the production and the emission layer. This conical-papillate shape is known to offer a vast surface area for evaporation and participate in the reflection of light. Osmophores and conical-papillate cells are typically responsible for releasing secondary metabolites into the atmosphere.

### 2.4.2 GLANDULAR TRICHOMES

*Trichomes* are hair-like structures on the surface of most plants, serving a number of functions ranging from protection against insect pests to heat and moisture conservation. Trichomes occur in plants in various forms and are sometimes very structurally complex (Peter & Shanower, 1998). *Plant glandular trichomes (GTs)* are adaptive structures that are well known as "phytochemical factories" due to their impressive capacity to biosynthesize and store large quantities of specialized natural products (Huchelmann et al., 2017; Y. Liu et al., 2019). Glandular trichomes (GTs) are anatomical structures specialized for the synthesis of secreted natural products. GTs are hairs on the epidermis and have cells specific to the biosynthesis and emission of abundant quantities of specific secretory products, such as nectar, mucilage, acyl lipids, digestive enzymes, or essential oils. These secreting trichomes are numerous and have very different morphologies in the plant kingdom. From our secondary metabolites' or essential oils' point of view, GTs are the sources that contain or secrete a mixture of organic compounds. GTs of plants are observed in detail using light, scanning, and transmission electron microscopy, reporting significant stages in the development of secretory cells, including their membrane system and nuclei, the overall size of the gland, and the amount of material released into the subcuticular cavity.

GTs are present in numerous monocotyledon plants of the *Tradescantia*, *Dioscorea*, and *Sisyrinchium* genera. GTs are more prevalent and are unique vegetative epidermal features of many families and genera, including the members of the *Lamiaceae, Asteraceae, Sphaerosepalaceae, Caryophyllaceae, Cucurbitaceae, Fabaceae, Rosaceae, Sapindaceae, Saxifragaceae, and Cannabaceae.* In the book *Anatomy of Dicotyledons* by Metcalfe and Chalk, the GT distribution list and morphological types in dicotyledons are provided. The diversity of secretory trichomes among the *Lamiaceae, Solanaceae* and *Rosaceaeisisis* are also captured and well-reported (Markus Lange & Turner, 2013; Metcalfe & Chalk, 1980).

Lange et al. have summarised studies on trichomes' current status and ongoing research. The isolation techniques for these trichomes for further studies are developed and reported for species like *Mentha x piperita, Mentha x Spicata, Helianthus annuus, Artemisia annua, Salvia officinalis, Thymus vulgaris, Rosa rugos* and so on. If the morphology is to be considered, peppermint showcases peltate GTs appearing evenly spaced and separated from each other by a similar number of epidermal cells, having predictable densities within different leaf regions (Markus Lange & Turner, 2013). GTs are in the form of modified epidermal hairs and are found covering leaves, stems, and parts of flowers in plants like lavender (Lavandula spp.), marjoram and oregano (Origanum spp.), and mint (Mentha spp.).

The structure of GTs indicates the secretory cell attachment by a single stem or basal cell in the epidermis. The outer surface is heavily cutinized, and a toughened cuticle usually completely covers the trichome. The metabolites are stored in subcuticular spaces and are released through the cuticle. The cells of GT have dense protoplasm that lacks a large central vacuole. There are numerous plasmodesmata (i.e., cytoplasmic threads running through cell walls, connecting the cytoplasm of adjacent cells) across the walls of the gland cells, especially between the stalk cell and the collecting cell.

## 2.4.3 Ducts and Cavities

Secreting cells such as ducts and cavities carry out the function of excreting gum, resin, paste or glue. Ducts and cavities are present in different plant families, such as *Apiaceae, Compositae, Rutaceae, Heliantheae,* and *Rubiaceae.*

*Secretory cavities* are more or less spherical structures. These are formed either from *parenchymas* or an *actual cell* disintegrating and leaving a cavity in the tissue. These spaces are lined with secretory cells, or an *epithelium*, that produces the essential oils. In high oil-yielding plants, several layers of these secretory cells are formed. The cavities continually enlarge, and some become filled with cells with thin, convoluted walls that store the oil produced from within their plastids (a class of cytoplasmic organelles). The simplest secretory structure is a single secretion-containing cell that can easily be distinguished from the adjacent non-secretory cells. Sometimes it is larger than the other cells or has a thick cuticularised lining. These cells can be found in different plant tissues like the *rhizome pith and cortex* of ginger (*Zingiber officinale Roscoe, Zingiberaceae*) and the *perisperm* and *embryo* of nutmeg (*Myristica fragrans Houtt., Myristicaceae*).

*Secretory ducts* are elongated cavities. They can often branch to create a network extending from the roots through the stem to the leaves, flowers and fruits. They are composed of an epithelium that surrounds a central cavity; this may result from *parenchyma* undergoing asynchronous divisions resulting in the expansion and formation of a cavity. Some cells forming the cavity wall will change into secretory epithelial cells. The oils are biosynthesized within their leucoplasts and move via the endoplasmic reticulum into the cavity. These cavities then become joined to form ducts. They can be found in all of the families *Apiaceae (Umbelliferae), Asteraceae (Compositae), Clusiaceae (Hypericaceae)* and *Pinaceae*. Numerous plants emit volatile organic compounds (VOCs) by non-specialized cells. Volatile monoterpenoids and sesquiterpenoids are emitted from the green leaves of such plants directly or after injury (Rehman & Asif Hanif, 2016).

## 2.4.4 Epidermal Cells

Glandular hairs do not usually secrete essential oils obtained from flowers but merely diffuse through the cytoplasm, the cell walls and the cuticle to the outside. The yield of essential oils from these species is generally very low. Examples include rose (Rosa spp., Rosaceae), 0.075% (w/v), acacia (Acacia spp., Fabaceae) 0.084% (w/v) and jasmine (Jasminum spp., Oleaceae) 0.04% (w/v). Buds of numerous plant genera, such as *Aesculus (Hippocastanaceae), Alnus (Betulaceae), Betula (Betulaceae), Populus (Salicaceae), Prunus (Rosaceae),* and *Rhamnus (Rhamnaceae),* also secrete lipophilic substances, mainly flavonoid aglycones mixed with essential oils. Secretion here occurs from epidermal cells that are covered by a cuticle. The secreted material is first eliminated into a space between the cells' outer walls and the cuticle covering them, forming a blister that subsequently bursts.

## 2.5    WATER IN TISSUE AND ITS FUNCTIONS

Having a brief insight into the structure of aromatic plant tissues, it is now important to understand the role, fundamentals, localization, and chemical state of water before undergoing the dehydration process. The water in the plant cells can be seen at four levels: *cellular level, sub-cellular level, molecular level,* and *plasticizers.*

In every cell, a plasma layer (cell membrane) separates the internal of the cell from the outside environment. Outside this membrane layer, the cell walls are present. Both these layers function differently. The membrane allows water to pass more quickly than the solute molecules. Cell walls readily pass both the solute and water molecules. Due to the presence of this membrane, osmosis is possible. With the help of osmotically active vacuolar solutes, pressure is exerted from the inside of the cell walls, keeping continuous elastic stress. This is called *turgor pressure,* which is responsible for the firmness and crispness of the tissues.

The role of water at the molecular level is to meet the integrity of membranes. Membranes are bilayers of lipids in the form of a liquid-crystalline state, having hydrated phospholipids by hydrogen bonding. This lipid bilayer structure is only possible in a hydrated medium. Water molecules positioned at the polar head of the membrane lipid form the hydrogen bond and strengthen the membrane structure. In the presence of sugars, water concentration lowers. This replaces sugar molecules with water molecules, protecting the membrane and avoiding structural collapse.

*Plasticizers* are lower molecular weight compounds, which lubricate the higher molecular weight polymeric compounds providing local mobility. Owing to these, glassy polymers turn rubbery. The water molecules play the role of plasticizer in the biopolymer. They act as mobility enhancers by increasing the free volume and lowering the viscosity. Water molecules can dramatically decrease the glass transition temperatures ($Tg$ of hemicellulose is lowered from 200°C to -10°C with a moisture content of about 30%) or act as an anti-plasticizer and lead to increased hardness (water activity aw = 0.1).

Water also acts as a transport fluid in all cells. This transfer of water molecules occurs in three modes. The movement within the extracellular spaces, external to the cell membrane, is termed *apoplastic transport.* The transfer between neighbouring cells through cytoplasmic strands is internal to the cell membrane and is termed *symplastic transport.* And the mass flow across the membrane is termed *transmembrane flux.* Hence, during drying, the plant tissues undergo *intercellular flux* (indicating water through cell walls), *wall-to-wall flux* (indicating capillary flux through cell walls) and *cell-to-cell flux* (indicating liquid water through vacuoles, cytoplasm and cell membranes) (Le Maguer et al., 2003; Prothon et al., 2003).

## 2.6 CONCLUSION

Plants comprise roots, rhizomes, stems, leaves, wood, flowers, seeds and fruits. These are built with cells and tissues in different arrangements with different functions. The complexity of the structures increases with the roles and functions being performed by the parts. In the case of aromatic plants, the secondary metabolites are secreted through the special tissues consisting of osmophores, glandular tissues, ducts and cavities. Depending on the genera of plants and the function of the secondary metabolites, the occurrence is seen in different parts; hence, the surrounding of this tissue is by a different type of cells.

Considering our interests, mass transfer pathways must be defined in the plant histology to preserve the secondary metabolites (in removal of water) or separate the secondary metabolites from the natural matrix (extraction of secondary metabolites). The plant structures are naturally defined pathways for transferring fluids, nutrients, gases, and so on, which can assist in the desired operations of drying and extraction if well studied. Before taking into account any protocol for dehydration or extraction, it is crucial to understand the cells and tissues, their chemical compositions, arrangements and functions before being subjected to the operation. This will help optimize the protocol and lead to the higher efficiency of the overall process.

## REFERENCES

Antoń, S., Kamińska, M., & Stpiczyńska, M. (2012). Comparative structure of the osmophores in the flowers of Stanhopea graveolens Lindley and Cycnoches chlorochilon Klotzsch (Orchidaceae). *Acta Agrobotanica*, *65*(2), 11–22. 10.5586/aa.2012.054

Carpita, N. C. (1996). Structure and biogenesis of the cell walls of grasses. *Annual Review of Plant Physiology and Plant Molecular Biology*, *47*(1), 445–476. 10.1146/annurev.arplant. 47.1.445

Curry, K. J., McDowell, L. M., Judd, W. S., & Stern, W. L. (1991). Osmophores, floral features, and systematics of Stanhopea (Orchidaceae). *American Journal of Botany*, *78*(5), 610–623. 10.2307/2445082

Effmert, U., Große, J., Rose, U. S. R., Ehrig, F., Kagi, R., & Piechulla, B. (2005). Volatile composition, emission pattern, and localization of floral scent emission in MIRABILIS JALAPA (NYCTAGINACEAE). *American Journal of Botany*, *92*(1), 2–12.

Huchelmann, A., Boutry, M., & Hachez, C. (2017). Plant glandular trichomes: Natural cell factories of high biotechnological interest. *Plant Physiology*, *175*(1), 6–22. 10.1104/pp.17.00727

Hülskamp, M., & Schnittger, A. (2012). Plant tissues. *ELS*, 1–5. 10.1002/9780470015902. a0002070.pub2

Le Maguer, M., Shi, J., & Fernandez, C. (2003). Mass transfer behavior of plant tissues during osmotic dehydration. *Food Science and Technology International*, *9*(3), 187–192. 10.1177/1082013203035392

Liu, H., & Specht, C. D. (n.d.). *Morphological Anatomy of Leaf and Rhizome in Zingiber officinale Roscoe, with Emphasis on Secretory Structures*. 1–4. 10.21273/HORTSCI14555-19

Liu, Y., Jing, S. X., Luo, S. H., & Li, S. H. (2019). Non-volatile natural products in plant glandular trichomes: Chemistry, biological activities and biosynthesis. *Natural Product Reports*, *36*(4), 626–665. 10.1039/c8np00077h

Markus Lange, B., & Turner, G. W. (2013). Terpenoid biosynthesis in trichomes-current status and future opportunities. *Plant Biotechnology Journal, 11*(1), 2–22. 10.1111/j.1467-7652.2012.00737.x

Megías, M., Molist, P., & Pombal, M. (2022). *Atlas of Plant and Animal Histology.* https://doi.org/http://mmegias.webs.uvigo.es/index.html

Metcalfe, C. R., & Chalk, L. (1980). *Anatomy of Dicotyledons - Volume I: Systematic Anatomy of Leaf and Stem, with a Brief History of the Subject* (Second). Oxford University Press.

Peter, A. J., & Shanower, T. G. (1998). Plant glandular Trichomes - Chemical Factories with Many Potential Uses. *Plant Glandular Trichomes, March*, 41–45. https://link.springer.com/content/pdf/10.1007%2FBF02837613.pdf

Prothon, F., Ahrne, L., & Sjoholm, I. (2003). Mechanisms and prevention of plant tissue collapse during dehdration - A Critical review. *Critical Reviews in Food Sci, 43*(4), 447–479.

Rehman, R., & Asif Hanif, M. (2016). Biosynthetic Factories of Essential Oils: The Aromatic Plants. *Natural Products Chemistry & Research, 04*(04). 10.4172/2329-6836.1000227

Weryszko-Chmielewska, E., & Chwil, M. (2012). Ecological adaptations of the floral structures of Galanthus nivalis L. *Acta Agrobotanica, 63*(2), 41–49. 10.5586/aa.2010.031

# 3 Need for Systematic/ Controlled Dehydration of Aromatic Plants

## 3.1 INTRODUCTION

Drying is a unique process for preserving agricultural produces, fruits, vegetables, medicinal herbs, and aromatic plants. Dehydration can also be considered an indispensable technique for large-scale agricultural production preservation. Dehydration is the removal of the majority of water contained in food items. From a unit operation point of view, drying includes transporting water molecules from the product matrix to the surroundings by altering the ambient conditions. The advantages of drying are as the moisture content is lowered, the microbial and enzymatic activities are reduced, increasing the shelf life. Along with improved shelf life, it reduces density and decreases transport costs.

Drying is the oldest food preservation technique known to human beings and includes sun drying or artificial dehydration of fruits, vegetables and oilseeds. The main purpose of drying products is to achieve longer periods of storage, lower the cost of packing and transportation, inhibit the microbial growth that causes decay spoilage and facilitate the formulation of product mixing for retailing (Murthy & Joshi, 2007). The goal of dehydration is to reduce the moisture content in the product by removal of the water by careful application of heat. A variety of processes can obtain dried or dehydrated agricultural products. These processes differ primarily by the drying method used, which depends on the type of food and the expected characteristics of the final product. The dehydration process needs to be suitably selected from amongst the existing technologies so that the water activity is adjusted to a level where microbial activity is least from a preservation point of view (Khaing Hnin et al., 2019). The demand for high-quality dried agricultural products is increasing worldwide (Chou et al., 2000). The quality of the dehydrated products can be mainly linked to the rehydration characteristics of the products (Savitha et al., 2022).

Herbs and spices have been known to be used for over 2,000 years (reported citations ~800–1000 BC). These were considered valuable products and were reported to be traded as commodities for their significance. The reported traditional applications of such herbs and spices were for their known medicinal properties, as preservatives – considering their powerful anti-oxidant properties, and for contributing flavour, aroma, and colour to culinary applications.

The dehydration process, specific to aromatic plant materials, should be able to retain the characteristics of the product from the application point of view (Thamkaew et al., 2021). Low-moisture products typically have a moisture content of < 25% and

DOI: 10.1201/9781003315384-3

water activity between 0.0 and 0.60; intermittent-moisture products have a moisture content between 15% and 50%; and water activity between 0.60 and 0.85 (Taoukis & Richardson, 2007). The most desired purpose of dehydration is to reduce moisture content in herbs, spices or other parts of aromatic and medicinal plants without affecting their key attributes for further use (Bhaskara Rao & Murugan, 2021).

## 3.2 HERBS, SPICES AND MEDICINAL PLANTS

### 3.2.1 CLASSIFICATION OF HERBS, SPICES AND MEDICINAL PLANTS

An *herb,* in botanical terms, is "any plant with soft succulent tissue," and as seen in the previous chapter, aromatic and medicinal plants contain secreted secondary metabolites in these soft and succulent tissues. Herbs can be classified as leafy products, and spices come as any part of the plant, like bud, bark, flower, fruit/berry, root/rhizomes, or seed (Pearson & Gillett, 1996). Based on the occurrence, these are sometimes also classified as "temperate zone origin" and "tropical aromatic," respectively. Though this classification is not clearly defined, key characteristics of these medicinal or aromatic herbs and spices are due to the presence of volatile oils and oleoresins in their parts.

Essential oils are extracted from several genera and are characterized in a small number of families – Lauraceae, Lamiaceae, Asteraceae, Myrtaceae, Rutaceae, Cuppressaceae, Poaceae, and Piperaceae – from different parts like peels, barks, leaves, flowers, buds, seeds, and others.

Herbs can be classified in three main ways – according to their usage, active ingredients, and period of life.

#### 3.2.1.1 Usage of the Herbs

1. *Medicinal herbs* – having curative powers and hence are used in making medicines for healing power.
2. *Culinary herbs* – having strong flavours and hence being used in cooking.
3. *Ornamental herbs* – having bright colours, looks, and unique textures and hence used for decoration.

#### 3.2.1.2 Chemical Compositions of the Herbs

1. *Aromatic (containing volatile oils)* – These have pleasant odours and are extensively used in therapeutic, flavouring and perfumery applications. These are further classified as *stimulant herbs,* which increase energy and activities of the body, or its parts or organs, and most often affect the respiratory, digestive, and circulatory systems.
2. *Nervine Herbs* – These are often used to heal and soothe the nervous system and often affect the respiratory, digestive, and circulatory systems as well.
3. *Astringents (containing tannins)* – These have the ability to precipitate the proteins, resulting in tightening, contraction or toning of living tissues and helping halt discharge. These often affect the digestive, urinary and circulating systems and act as analgesic, antiseptic, antiabortive, astringent, emmenagogue, homostatic, and styptic.

4. *Bitter* – These contain phenolic compounds, saponins, and alkaloids and are further classified as *laxative herbs* (include alterative, anticatarrhal, anti-pyretic, cholagogue, purgative, hepatonic, sialagogue, vermifuge, and blood purifier by nature), *diuretic herbs* (alterative, antibiotic, anticatarrhal, anti-pyretic, antiseptic, lithotriptic, and blood purifier in nature), *saponin-containing herbs* (known for their ability to produce frothing or foaming and emulsifying fat-soluble molecules in the digestive tract to enhance the body's ability to absorb other active compounds), and *alkaloid-containing herbs.*

5. *Mucilagnious (containing polysaccharides)* – These have a slippery, mildly sweet taste in water and contain mucilage, which is not broken down by the human digestive system but absorbs toxins from the bowel and gives bulk to the stool. They are effective topically as poultices and knitting agents, and for demulcent action on the throat when used in lozenges or extracts. They are antibiotic, antacid, demulcent, emollient, culnerary, and detoxifiers in nature.

6. *Nutritive (for foodstuffs)* – These are true foods and provide some medicinal effects such as fibre, mucilage, and diuretic action. Most importantly, they provide the protein, carbohydrates, and fats, plus the vitamins and minerals that are necessary for adequate nutrition.

### 3.2.1.3 Period of Life

1. *Annual herbs* are the ones that complete their life cycle in 1 year, starting from seeds.
2. *Perennial herbs* are the ones that grow for more than one season.
3. *Biennial herbs* are the ones that live for two seasons and bloom in the second season.

Medicinal and aromatic plants can also be classified (Embuscado, 2015; Lawless, 1992; Opara & Chohan, 2014; Pandey et al., 2019; Vázquez-Fresno et al., 2019) based on

- *Botanical taxonomy* (Embryophyta – Gymnospermae & Angiospermae),
- *Growth Habitat* (herbs – 32%, shrubs – 20%, climbers – 12%, trees – 33% and liners – 3%),
- *Modes of nutrition* - (autotroph, symbiotic and heterotrophs),
- *Habitat* – (tropical, sub-tropical and temperate).

Considering the vast classification, applications and significance of these medicinal and aromatic herbs, it is important to consider each plant uniquely. Moreover, considering the lifecycle of each herb, active ingredients associated, means of administration and applications, it is important to process the cultivated harvest either through dehydration or extraction. Since these secondary metabolites are in small quantities in the plant material, huge quantities of herbage must be processed during harvest, considering the low shelf life and the industrial demand. These constraints, like various herbs, selective applications, period and season of harvesting, chemical compounds of interest from different herbs and volume of harvested plants for handling, bring out the criticalities and attention on the need for dehydration and preservation of the biomasses. Dehydration is the best way to preserve the natural volatiles in their natural matrix. It is also a prerequisite for carrying out the extraction of the marker compounds, making dehydration a critical operation (Vlaic et al., 2022).

### 3.2.2 SHELF LIFE OF AROMATIC PLANTS AND MEDICINAL HERBS

The ideal time to harvest most herbs is just before the flowers first open, when they are in the bursting bud stage. It is the point at which the plants have reached full maturity, and the secreted secondary metabolites are still stored in the secretory glands, ducts or cavities and yet to be released. Gathering the herbs is highly recommended in the early morning hours (mostly for flowers having delicate tissues) after the dew has evaporated to minimize wilting. Harvested herbs are required to be processed timely to preserve their quality and organoleptic characteristics. After harvesting, herbs need to be processed or preserved in their balsamic moments (richer in active ingredients and fragrances).

Whether vegetables, aromatic plants or medicinal herbs or fruits, fresh-cut and minimally processed plant materials are known to have the highest nutritional value. A segment of the food processing industry related to fresh-cut agro-produces is continuously evolving and pursuing innovations. Fresh-cut medicinal and aromatic herbs, also being a part of that domain, are gaining popularity owing to their intense flavour and convenience. These are to be minimally processed (washed, cut and packaged at chilling temperatures) to assure the safety and maintenance of their freshness, tenderness and uniformity of colour for a longer period. Fresh aromatic herbs clearly have advantages over the dried product in the sense of more retained aromatic compounds. Minimal processing of herbs results in an increase in the respiratory rate, leading to a more prompt onset of senescence signs and, consequently, loss of quality. Though fresh-cut herbs are advisable, quality studies of aromatic herbs indicate that stable phytonutrient contents start changing after a storage duration of 10 days, giving a low shelf life and processing time. Hence, it is recommended to process these herbs to meet the desired form of retention or extract these phytonutrients to the maximum in this short shelf life duration (Santos et al., 2014).

Water being a significant component of plant material, the first step involved in a post-harvest operation is drying the crop. Medicinal and aromatic plants are cultivated for the extraction of active ingredients. There are significant influences of agro-ecological conditions on the active ingredients, whether from wild gathering or cultivation. Considering the small content of active ingredients/secondary metabolites (0.1%–2%), centralized processing for the harvested/collected herbs is mandatory to standardize the composition of desired extracts (Oztekin & Martinov, 2007). Centralized processing adds to the requirement of the transportation, storage and processing of high herbage volumes in the limited time span equivalent to that of the shelf life for fresh-cut herbs. If the water content in the plant material is lowered or if the water activity is met, then the shelf life increases for the product, providing more time span for the processing of the harvest and lowering the processing capacity of the centralized processing plant (Figure 3.1).

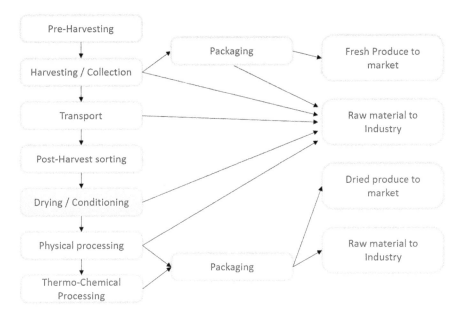

**FIGURE 3.1** Flowchart for aromatic plants from cultivation to consumption.

## 3.3 IMPORTANCE AND NEED OF CONTROLLED DEHYDRATION

In medicinal and aromatic plants, drying becomes the fundamental unit operation for preserving plant materials and active ingredients in the matrix post-harvesting. The bio-origin products need to be dried as a pre-treatment before subjecting to extraction. Low-temperature drying operations are typically recommended, though this process is associated with higher processing time and low energy efficiency, considering the thermal instability of active ingredients. Owing to the temperature sensitivity of the active ingredients and the mandatory need for the drying process, it plays an authoritative role in the product quality and hence the value of the products. This brings the need for a literature review for the optimal drying techniques, operational protocol, optimization of parameters, experimental trial runs and critical scale-up studies for optimizing utility requirements and meeting product quality.

Nine factors can be controlled to meet the drying kinetic requirement described in the later chapter. To meet the industrial drying capacity, activation energy/drying temperature should be as high as possible without reducing the quality of the product below the threshold of the buyers' requirements. Such maximum temperatures permissible for the dehydration of various medicinal, aromatic and spices plants vary depending on the active ingredients of the species. Aromatic herbs may be said to be more vulnerable to higher temperatures than spices and medicinal herbs based on the natural volatiles and chemicals associated as key ingredients. It is reported that almost 90% and about 50% of the essential oils may be lost when completely dried at 90°C and 60°C for *Salvia officinalis*. A similar study for *Piper betel* indicated that retention of volatile oils in dried piper betel varied between 9.5% and 8.75% (w/w) when dried at

40°C and 60°C under controlled conditions. This indicates that almost 8%–10% of volatile oils may get lost when dried at temperatures in the mid-range (Müller, 2007; Pise et al., 2022). These studies also indicate that the volatile oil loss significantly increased as the moisture content was reduced below 50% (wet basis), indicating not only control over external conditions but also retention of low water contents is required to preserve the key ingredients in the natural matrix (Table 3.1).

**TABLE 3.1**
**Loss of Active Ingredient During Convective Drying (Müller, 2007)**

| Species | Active Ingredients | Temperature of Drying Air (°C) | | | | | | | | | | |
|---|---|---|---|---|---|---|---|---|---|---|---|---|
| | | Fresh | 30 | 35 | 40 | 45 | 50 | 60 | 70 | 80 | 90 | 100 |
| Aloe barbadensis | Acemannan | 0* | 11 | – | 14 | – | 14 | 16 | 23 | 28 | – | – |
| Artemisia dracunculus | Essential oil | 0* | – | – | – | 63 | – | 75 | – | – | 36 | – |
| Artemisia dracunculus | Essential oil | 0* | – | – | – | 75 | – | 91 | – | – | 74 | – |
| Artemisia dracunculus | Essential oil | – | – | – | 0* | – | 40 | 40 | 60 | 60 | 60 | 80 |
| Chamomilla recutita | Essential oil | 0* | 14 | – | 12 | – | 24 | 21 | 21 | 24 | 19 | – |
| Cymbopogon citratus | Essential oil | – | 0* | – | – | – | –7 | – | 11 | – | 21 | – |
| Echinacea angustifolia | Echinacoside | 0* | 29 | – | 31 | – | 38 | 45 | – | – | – | – |
| Echinacea purpurea | Alkylamides | – | 100 | – | 0* | – | – | 20 | – | 18 | – | – |
| Hypericum perforatum | Hypericin | – | 0* | – | 17 | 13 | 15 | –14 | 14 | –23 | 12 | – |
| Lipia alba | Essential oil | 0* | – | – | 19 | – | 15 | 15 | 15 | 15 | – | – |
| Mentha x piperita | Essential oil | 0* | – | – | – | – | 84 | – | 98 | – | – | – |
| Ocimum basilicum | Essential oil | – | – | – | 0* | – | 30 | 30 | 30 | 70 | 70 | 90 |
| Origanum majorana | Essential oil | – | – | – | 0* | – | 24 | 24 | 41 | 41 | 59 | 71 |
| Origanum vulgare | Essential oil | 0* | – | 14 | – | 9 | – | – | – | – | – | – |
| Rosmarinus officinalis | Essential oil | – | – | – | 0* | – | 13 | 13 | 26 | 26 | 57 | 87 |
| Salvia officinalis | Essential oil | 0* | –1 | – | – | – | – | 31 | – | – | – | – |
| Salvia officinalis | Essential oil | – | – | – | 0* | – | 15 | 50 | 75 | 75 | 80 | 85 |
| Salvia officinalis | Essential oil | 0* | 0 | – | 0 | – | 0 | 20 | 48 | 63 | 80 | – |
| Saturea hortensis | Essential oil | – | – | – | 0* | – | 23 | 38 | 46 | 62 | 98 | 100 |
| Tanacetum parthenium | Parthenolide | 0* | – | – | –12 | – | – | –2 | 10 | 25 | 22 | – |
| Thymus vulgaris | Essential oil | 0* | 1 | – | – | – | – | 43 | – | – | – | – |
| Thymus vulgaris | Essential oil | – | – | – | 0* | – | 0 | 0 | 70 | 70 | 80 | 100 |

## 3.4   ROLE OF DRYING IN EXTRACTION

Medicinal and aromatic plants, typically flowers, fruits, leaves and herbs, contain 70%–95% of water, thus limiting their shelf life. Extending shelf life for storage, transport, processing, and better utilization is crucial in these plant materials. Pre-treatments, like sorting, size reduction, cleaning, washing and separation of selective parts, before drying or extraction may be applied, but they are specific to each plant and dependent on the final objective. Some drying methods, such as freeze-drying, do not use temperature to dry the material; therefore, it can avoid biomass degradation. However, it may lead to the degradation of fibre structures and hence affect rehydration. Mild drying conditions may not degrade the constituents of medicinal and aromatic plants and are recommended under lower pressure conditions. Some other pre-treatments, even those using higher temperatures, could be eventually used in cases where they do not affect the compounds of interest.

Nevertheless, drying and grinding are the usual conditioning stages of plants and herbs subjected to supercritical fluid extraction using $CO_2$ or solvent extraction. Sample drying is usually required to facilitate stable storage and to facilitate extraction with a pure solvent to have control over the polarity. Very rarely, the increase in moisture content increases the solubility of essential oil, lowering the solvent requirement, as reported in the case of Helichrysum italicum flowers (from 10.5% to 28.4%) (Ivanovic et al., 2011). Similarly, sometimes, preheating as a part of dehydration operation at elevated temperatures just before extraction is recommended to increase the yield as reported for cannabinoids (Ribeiro Grijó et al., 2019).

In the case of extraction, reported literature also states that particle size reduction affects the internal mass transfer and hence the quality of the oil because of the varying content of waxes. The preprocess of particle size reduction exposes the internal tissue structures for increased surface contact area of the matrix and the solvent. Drying helps make the plant material brittle, easy to crush or crumble, and reduces particle sizes. Having said so, control over dehydration helps control brittleness, which could help obtain optimal particle size. Hence, drying is important for reducing water activity, controlling microbial activity and increasing shelf life, and it also plays a role in pre-treatment for optimization of extraction (López-Hortas et al., 2022).

## 3.5   CONCLUSION

The complexity of the plant-based aroma and essential oils, considering the chemical composition, is well recognized. Since selective compounds in the oil plays a critical role in the respective application, extraction of the same in primitive form is crucial. To process these effectively, centralized processing is required to maintain the consistency of quality. The need for transportation, handling & storage, and timely processing of seasonal high-volume harvest effectively and efficiently brings out the significance of dehydration operation. Most aromatic plants are temperature sensitive, so the dehydration conditions need to be appropriately selected to retain the maximum amount of the active ingredient.

## REFERENCES

Bhaskara Rao, T. S. S., & Murugan, S. (2021). Solar drying of medicinal herbs: A review. *Solar Energy*, *223*(June), 415–436. 10.1016/j.solener.2021.05.065

Chou, S. K., Chua, K. J., Mujumdar, A. S., Hawlader, M. N. A., & Ho, J. C. (2000). On the intermittent drying of an agricultural product. *Food and Bioproducts Processing: Transactions of the Institution of of Chemical Engineers, Part C*, *78*(4), 193–203. 10.1205/096030800051065296

Embuscado, M. E. (2015). Spices and herbs: Natural sources of anti-oxidants - A mini review. *Journal of Functional Foods*, *18*, 811–819. 10.1016/j.jff.2015.03.005

Ivanovic, J., Ristic, M., & Skala, D. (2011). Supercritical CO2 extraction of Helichrysum italicum: Influence of CO2 density and moisture content of plant material. *Journal of Supercritical Fluids*, *57*(2), 129–136. 10.1016/j.supflu.2011.02.013

Khaing Hnin, K., Zhang, M., Mujumdar, A. S., & Zhu, Y. (2019). Emerging food drying technologies with energy-saving characteristics: A review. *Drying Technology*, *37*(12), 1465–1480. 10.1080/07373937.2018.1510417

Lawless, J. (1992). The Encyclopedia of Essential oils-Complete guide for the use of Aromatic oils in Aromatherapy, Herbalism, Health and well-being. In *Conari Press*. Haper Collins.

López-Hortas, L., Rodríguez, P., Díaz-Reinoso, B., Gaspar, M. C., de Sousa, H. C., Braga, M. E. M., & Domínguez, H. (2022). Supercritical fluid extraction as a suitable technology to recover bioactive compounds from flowers. *Journal of Supercritical Fluids*, *188*(January). 10.1016/j.supflu.2022.105652

Müller, J. (2007). Convective drying of medicinal, aromatic and spice plants: A review. *Stewart Post-harvest Review*, *3*(4). 10.2212/spr.2007.4.2

Murthy, Z. V. P., & Joshi, D. (2007). Fluidized bed drying of aonla (Emblica officinalis). *Drying Technology*, *25*(5), 883–889. 10.1080/07373930701370290

Opara, E. I., & Chohan, M. (2014). Culinary herbs and spices: Their bioactive properties, the contribution of polyphenols and the challenges in deducing their true health benefits. *International Journal of Molecular Sciences*, *15*(10), 19183–19202. 10.3390/ijms151 019183

Oztekin, S., & Martinov, M. (2007). *Medicinal and Aromatic Crops - Harvesting, Drying, and Processing* (S. Oztekin & M. Martinov (eds.)). Haworth Forr and Agricultural product Press.

Pandey, A. K., Kumar, P., Saxena, M. J., & Maurya, P. (2019). Distribution of aromatic plants in the world and their properties. In *Feed Additives: Aromatic Plants and Herbs in Animal Nutrition and Health* (pp. 89–114). Elsevier Inc. 10.1016/B978-0-12-814700-9.00006-6

Pearson, A. M., & Gillett, T. A. (1996). Herbs, Spices, and Condiments. In: Processed Meats. In *Journal of AOAC INTERNATIONAL* (Vol. 79, Issue 1, pp. 199–199). Springer, Boston, MA. 10.1093/jaoac/79.1.199

Pise, V., Shirkole, S., & Thorat, B. N. (2022). Visualization of oil cells and preservation during drying of Betel Leaf (Piper betle) using Hot-stage Microscopy. *Drying Technology*. 10.1080/07373937.2022.2048848

Ribeiro Grijó, D., Vieitez Osorio, I. A., & Cardozo-Filho, L. (2019). Supercritical Extraction Strategies Using CO2 and Ethanol to Obtain Cannabinoid Compounds from Cannabis Hybrid Flowers. *Journal of CO2 Utilization*, *30*(January), 241–248. 10.1016/j.jcou.2018.12.014

Santos, J., Herrero, M., Mendiola, J. A., Oliva-Teles, M. T., Ibáñez, E., Delerue-Matos, C., & Oliveira, M. B. P. P. (2014). Fresh-cut aromatic herbs: Nutritional quality stability during shelflife. *Lwt*, *59*(1), 101–107. 10.1016/j.lwt.2014.05.019

Savitha, S., Chakraborty, S., & Thorat, B. N. (2022). Microstructural changes in blanched, dehydrated, and rehydrated onion. *Drying Technology*, *40*(12), 2550–2567. 10.1080/07373937.2022.2078347

Taoukis, P. S., & Richardson, M. (2007). Priciples of Intermediate Moisture foods and related Technology. In *Water Activity in Food - Fundamentals and applications* (pp. 273–313). Blackwell Publishing.

Thamkaew, G., Sjöholm, I., & Galindo, F. G. (2021). A review of drying methods for improving the quality of dried herbs. *Critical Reviews in Food Science and Nutrition*, *61*(11), 1763–1786. 10.1080/10408398.2020.1765309

Vázquez-Fresno, R., Rosana, A. R. R., Sajed, T., Onookome-Okome, T., Wishart, N. A., & Wishart, D. S. (2019). Herbs and Spices- Biomarkers of Intake Based on Human Intervention Studies - A Systematic Review. *Genes and Nutrition*, *14*(1), 1–27. 10.1186/s12263-019-0636-8

Vlaic, R. A. M., Mureşan, V., Mureşan, A. E., Mureşan, C. C., Tanislav, A. E., Puşcaş, A., Petruţ, G. S. M., & Ungur, R. A. (2022). Spicy and Aromatic Plants for Meat and Meat Analogues Applications. *Plants*, *11*(7), 1–21. 10.3390/plants11070960

# 4 Drying Technology and Selection of Dryers

## 4.1 INTRODUCTION

*Drying*, by definition, is *"the separation operation that converts solids, semisolids or liquids to solid products by evaporation of liquid into the vapour phase through the application of heat."* In the case of freeze-drying, removal occurs by sublimation below the triple point, i.e., solid phase, directly to the vapour phase. Drying is one of the oldest, most common and most diverse unit operations in chemical engineering. Though extensively researched, and with several technologies meeting maturity, innovative approaches are still required in the domain of drying to meet new challenges, global competitiveness, energy efficiency, environmental impacts and meeting desired product qualities.

Being associated with the latent heat of vaporization, it is one of the most energy-intensive and inherently inefficient unit operations similar to distillation. Drying is a mandatory unit operation for one or several reasons such as ease of handling, preservation, storage, transport cost reduction, achieving desired product quality, and so on. Having said so, drying is also a critical and systematic process to be conducted with proper understanding, as improper drying may lead to irreversible damage to quality, making the product lose value.

Drying can be considered as *"an art of living"* as it had been a part of life (as far as 20000 BC) for drying the preservation of foods for ages. This sort of drying was dependent on the sun and influenced by weather, causing a lack of product quality control. During natural drying, the desiccating effect of air provides the mass transfer and activation energy for evaporation of water molecules by exposure to the sun. In recent years technology has been developed for controlling the exposure and enhancing the utilization of solar energy through various configurations of solar dryers (Chavan et al., 2020). Industrial-scale drying of different products leads to the implementation of mechanical drying to cater to the capacity in the specified time (Hayashi, 1989). Some mechanical drying technology developed to meet industrial demands include drum drying, trucked tray dryers, flash drying, fluidized drying, vacuum drying, spray drying and freeze-drying, which are well-researched and reported.

## 4.2 FUNDAMENTALS OF DRYING

### 4.2.1 DRYING PROCESS AND IMPACTING PARAMETERS

Drying occurs by the vaporization of liquids through the activation of water molecules/supply of heat followed by the provision of the diffusion pathway to transfer

DOI: 10.1201/9781003315384-4

vapours. The heat supply can be through *convection, conduction* (in the mode of direct or indirect contact), or *radiation* through various frequencies. The temperatures are maintained in the drying chamber from freezing to hot conditions, i.e., −50°C to 100°C for herbs and spices, using refrigerated systems and heaters (fuelled by electricity, natural gas, coal/briquettes, firewood and so on), respectively. Considering the size, shape and structure of the objects, this temperature or the activation energy can be supplied through direct or indirect heat, such as conduction, convection or radiation mode (microwave, far-infrared, infrared, ultraviolet and radio-frequency energy) of heat transfer.

The diffusion of vapours is considered in two-step internal diffusion (movement of vapours in the interiors of the matrix) and external diffusion (removal of the vapours from the surface), which basically govern the drying regimes (Arun S. Mujumdar & Devahastin, 2000; Pai et al., 2021). The external mass transfer conditions can be controlled by circulation draft (natural or forced), vacuum (regulating partial pressures) and relative humidity for providing moisture gradient. Regulating these parameters, the energy of activation for vaporization and the inherently slow process of drying makes this unit operation one of the most energy-intensive (Arun Sadashiv Mujumdar, 2014).

From a research point of view, the drying process mainly depends on the product size (microns to tens of centimetres), product porosity (0%–99%), drying time (0.25 secs to 5 months), production capacity (0.1 kg/h. to 100 tons/h.), product speed (stationary to 2,000 m/min) or the residence time, drying temperature (below triple point to above critical point of water), operating pressure (fraction of millibar to 25 atm) and mode of heat transfer (Arun Sadashiv Mujumdar, 2014). Thus, it can be seen that the drying process needs a great deal of attention. The right choice of the most suitable dryer must provide a good return on investment and be techno-economically feasible (Figure 4.1).

## 4.2.2 Drying/Dehydration Fundamentals

Drying is a complex operation having not just parallel heat and mass transfer occurring at different transfer rates but is also accompanied by physical and chemical transformation. These physical and chemical transformation interns affect the heat and mass transfer rate making it a dynamic phenomenon to be monitored.

*Physical* changes occurring during the process may include shrinkage, swelling, puffing, crystallization, glass transition, pore collapse and so on. *Chemical or biochemical* changes may be seen as colour/browning, texture, odour, protein denaturation, oxidation of lipids, destruction of vitamins, and so on.

During dehydration, moisture concentration changes in the plant material along with the physical structures, complicating the process. Reduced concentration gradient impacts the diffusion of moisture and hence the rate of drying. Hence, the moisture transfer rate may also change with the elapse of time for dehydration. These changes are encountered owing to the change in the thermodynamic properties of the air–water mixture and the moist plant material. The thermodynamic properties mainly accounted for monitoring the changes during the drying operation include *psychrometry, equilibrium moisture content, and water activity.*

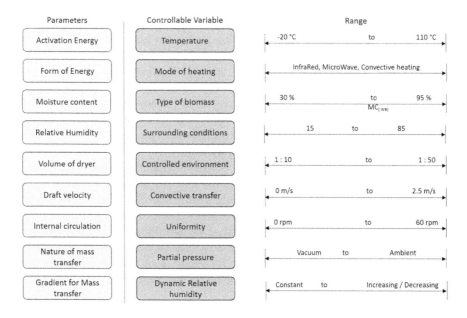

**FIGURE 4.1**   Parameters that can be considered for the selection of dryer and drying conditions.

*Psychrometry is the study of hygrothermal properties of humid air, indicating the relationship between temperature and absolute humidity represented on a psychrometric chart used for designing dryers.*

*Equilibrium moisture content* is the moisture content of the wet solid in equilibrium with the humidity of the surrounding air at that temperature indicated in terms of sorption isotherm. The equilibrium moisture content or the sorption isotherms are used to understand the binding mechanism of water molecules at individual sites.

*Water activity* is the ratio of the partial pressure of water over the wet solid system to the equilibrium vapour pressure of water at the same temperature. This indicates the availability of water in the system (dried product).

As stated above, these three hygrothermal properties, along with the physical properties and changes occurring in the material, determine the drying rate at a fixed condition known as the *drying kinetics*. These fundamentals and related terminologies are explained in detail by Mujumdar et al. and Kaur et al. (Barjinder et al., 2018; Arun S. Mujumdar & Devahastin, 2000).

## 4.3   DRYING OF AGRICULTURAL PRODUCTS

Drying is a unique process for preserving agricultural produces, fruits, vegetables, medicinal herbs, and aromatic plants. Drying lowers the moisture content reducing the microbial and enzymatic activity by lowering water activity. Most Asian countries majorly carry out agricultural product exports. These large quantities of produce are subjected to drying for maintaining the shelf life, improving taste, enhancing appearances and reducing transport costs. Seen as an industrial process, high-value products target controlled drying conditions to meet product quality, whereas low-value products target reducing utility consumption (Chou et al., 2000). Processing time is equally important for industrial-scale drying apart from product quality. This brings the focus and attention towards research and optimization of drying features like thermal energy saving, effective drying time, moisture rate removal, product surface temperatures and product quality.

In the case of agricultural products, considering the heat sensitivity of products, the drying requirement shifts from constant temperature drying to complex time-varying drying schemes. Considering the volume being subjected to drying and various challenges over physical, chemical and biochemical properties, a large assortment of dryers have been developed for meeting the drying, preservation and techno-economic feasibility needs. More than 500 dryers are reported in the literature for agricultural product dehydration, and around 100 types of dryers are commercialized. These differences in the designs are mainly seen due to the physical attributes affecting the drying parameters.

Parameters taken into consideration include the products being processed, modes of heat input, temperature & pressure of operation, quality specifications of the products required and so on. Typically, dryers are designed to consider hot air as medium, convection as the heat transfer mode, and atmospheric pressure operating conditions under steady operating conditions in continuous mode for bulk volumes. However, these are reported to showcase limitations like non-uniform product quality due to long or inadequate exposure to drying medium in a non-uniform manner. This indirectly results in non-uniform temperature profiles and drying rates, and hence non-uniform drying times, textures, and colours. The complexity of ongoing drying also results in changes in the product's physical, chemical, rheological and sensory attributes; therefore, extensive research to overcome operational problems and issues has been reported and is being continued (Arun S. Mujumdar & Law, 2010).

## 4.4 HISTORY AND DEVELOPMENT OF DRYERS

The change in drying factors was prominently seen from 1970 onwards. Until 1960, there were negligible variations seen in the dryers concerning the relative volume of the air intake and size of the chamber or air blast, and they were typically horizontal with a parallel flow or cylindrical vertical flow. Slowly, the developments were seen with tall vertical chambers, increased ventilation power, water evaporation control, volumetric heat transfer coefficient, thermal efficiency and evaporation capacity. The dryers were soon characterized based on *temperatures* – (–)40°C to 0°C (vacuum freeze dryers), 0°C to 80°C (vacuum, pneumatic, tray, rotary, spray and solar), and 120°C to 180°C (cylindrical, fluidized, rack, spray, flash dryers and so on) and *time* – seconds (spray and flash), *seconds to minutes* (rotary, tray, pneumatic), *minutes* (vacuum, fluidized, trays), *hours* (rack and freeze), and *days* (sun and solar). Further factors like the products and process, capacities, quality of products and quality control, environmental impacts, energy efficiency and recovery systems, the safety of operations, CAPEX and OPEX, time for processing and so on were also considered for the development of new dryers. Recently the focus has been shifted to multi-mode heat inputs, consideration for time-dependent heat inputs and drying kinetics, use of super-heated steam, sub-atmospheric pressures, low-temperature and dehumidified air, multistage and hybrid drying and so on (Arun S. Mujumdar & Law, 2010).

### 4.4.1 DEVELOPMENT OF TECHNOLOGY AND TYPES OF DRYERS

Drying of the product includes contacting air with the product at various temperatures for the mass transfer of moisture from the feed product to the surrounding atmosphere. These are dynamics of changes occurring for heat and mass transfer. The heating of the product is accompanied by the cooling of the air, and the reduction in moisture content of the product is accompanied by the increased relative humidity of the surrounding.

Typically, the drying unit operation falls under two methods – *natural and artificial drying. The natural drying method* is undertaken by direct or indirect exposure to the sun, especially when the relative humidity is below 70% and sufficient space is available for spreading out agricultural produces. *The artificial drying method* is a part of the progressive mechanization of agriculture. It supports the drying operation even in humid and subtropical zones with unfavourable conditions.

The construction of a mechanical dryer includes the chamber/body of the dryer, the heating element or activation energy source and the ventilators. Further, the construction can be categorized as *static dryers* or *continuous* dryers. Static dryers are inexpensive, low maintenance, and adaptable to small and marginal collection centres. Continuous dryers are high-flow dryers with complex constructions, complementary equipment, planning, and organization. These are more favourable or moreover required for high-volume harvest processing (Gunathilake et al., 2018).

Different types of dryers have been designed and implemented for agricultural produces, especially grains, considering their processing volumes. Certain types are listed below,

- *Batch dryer*
- *Batch-type continuous flow dryer*
- *Batch-type fluidized bed dryer*
- *Multi-crop dryer*
- *Continuous fluidized dryer*

These types may further be subjected to minor modifications to meet specific crop requirements.

The types of dryers that are defined based on the drying chambers and are used for perishable crops include,

- *Tray Dryers* – This is the simplest dryer in which trays are used for spreading out the product to be dried uniformly in thin layers.
- *Tunnel Dryers* – This is a developed version of tray dryer in which trolley trays move through the tunnel.

- *Pneumatic Dryers* – Products to be dried are rapidly conveyed in the air stream.
- *Rotary Dryers* – Horizontal, inclined cylinders are used for handling the material. Either the cylinder rotates, or the paddles or screws rotate and the cylinders are stationary.
- *Trough Dryers* – A trough-shaped belt is used for transport of material while air is blown through the bed of material.
- *Bin Dryer* – This has bins with perforated bottoms that are used for drying the products.
- *Belt Dryer* – This uses a horizontal mesh or solid belt for material transport and passing of the drying medium.

Different types of dryers developed based on different techniques used for meeting the final product requirement include,

- *Freeze Dryer* – As the name suggests, the process includes freezing the feed product to be dried at a very low temperature of $(-)80°C$ to $(-)50°C$. At these conditions, the water solidifies and then is removed in the form of water vapours by sublimation. Two major issues with this process are that it is ecologically unfriendly, and the fibres are damaged during drying, leading to brittle structures.
- *Spray Dryer* – This method was developed for obtaining dried powders from slurry by rapid evaporation. This method is widely used in industries for obtaining a wide range of products with low water activity, longer shelf life and good quality thermally sensitive feed products. Applications for obtaining dried pharmaceuticals, milk, fruit juices, prebiotics, and pro-biotics products are well established.
- *Flash Dryer* – Flash drying is similar to spray drying in which super-heated fluid/slurry is sprayed or exposed to atmospheric conditions in the agitated state so that the moisture content is released rapidly giving a dehydrated product.
- *Drum Dryer* – This is a method developed for processing pureed raw materials. The liquid slurries are dried at relatively low temperatures over high-capacity rotating drums collected in the form of sheets.
- *Vacuum Dryer* – The drying conditions in this process are changed to lower pressures by removing the air around the product to be dried. This leads to faster removal of moisture from the product at atmospheric temperatures. This process ensures taste, colour, nutritional values, and active ingredient retention. The fibre structures, in this case, are also retained, providing a good quality reconstituted product on rehydration. This is the most favoured process for drying thermally sensitive products like aro-matic agro-produces, including herbs and spices.
- *Microwave-Assisted Dryer* – Microwaves can be successfully adopted for drying food crops with the advantage of less drying time. Microwaves can penetrate moist food material vibrate water molecules, and generate heat. However, it is associated with the disadvantages of non-homogeneous

drying of the material. This can be overcome by clubbing with a fluidized bed to get homogeneous heating. It is also reported that microwave and fluidized bed have resulted in marked time and energy saving compared to conventional fluidized bed drying.

Various types of dryers have been researched, developed and applied for agricultural drying. Most agricultural products have significant moisture content during the harvesting stage. The key challenges encountered for agricultural products can be narrowed down to the irregular shape of the agricultural produces (grains, legumes, fruits & vegetables, classified under durable and perishable harvests), irregular matrix of the agricultural produce with complex tissues building blocks as seen in an earlier chapter, significant volumes for processing and short times for processing. During the drying operations, challenges like inhomogeneous drying (over- or under-dried parts), the efficiency of the process and the quality of the product (colour, texture, brittleness, taste, physical and chemical attributes) also need to be considered during the design of dryers. Selection of an appropriate dryer is crucial, especially when minor changes in the feed or final product characteristics could adversely affect the performance.

Recent industrial-scale agricultural cultivation and significant advances in research and technology have led to the exploration and implementation of new technologies for the dehydration of agricultural produces as well as processed foods. These are crucially developed to meet the well-controlled dehydration requirement and the product requirement as demanded by the market. Such advances have led to further different types of dryers, including,

- *Multi-Stage Dryer* – This method is developed based on the heat and mass transfer characteristics along with the drying curve showcased by the material. Keen observation of drying curves and heat and mass transfer rates indicate distinct mechanisms for optimized drying rates in different conditions. Ideally, each mechanism needs different operating conditions and different techniques for drying. This calls for the need for a multi-mode / multistage dryer. With such a dryer, it is possible to vary the temperature and the drying medium flowrates at different dryer cases or stages as required. Advantages seen are the utilization of low-temperature secondary energy flow in the dryer and integration with the existing system to improve efficiency.
- *Super-Heated Steam Dryer* – The replacement of hot air with super-heated steam as a drying medium was an idea dated long back but is now gaining momentum. It is a revolutionary technique considering drastic changes in operating conditions. The equipment, as such, needs auxiliary devices for meeting the heating and cooling of the product, and the chamber can either be pressurized or under vacuum. The concept behind using the super-heated steam is to increase the mobility of moisture in the feed material to increase the drying rate.
- *Contact-Sorption Dryer* – In this method, the feed material to be dried is contacted with heated particulate inert material. The particle-particle

contact induces heat transfer and hence mass transfer facilitating the drying operation. The heated, highly hygroscopic inert material generates a high chemical potential gradient. The drying rate is further enhanced in this method by the introduction of sorbent material along with hot air. This is also termed *two-phase drying*. Typically sand and silica gel are used for drying in this method.

- *Spouted-Bed Dryer* – These are modified fluidized beds for coarse particles that are large and heavier. The design is such that the motion is rather regular than random. These designs are further modified to have the rotating jet spout-bed dryers. Such fluidized bed dryers are highly recommended for heat-sensitive particles.

- *Heat Pump Dryer* – Heat pumps are known to be efficient for drying operations. It provides an advantage of the recovery of heat for the exhaust and control over temperature and mainly humidity of the drying medium. It can be stated to be a refrigeration system in reverse operation. The refrigeration cycle in this process is used for dehumidification of air and heating the air for sensible heating to the desired temperature. A well-designed heat pump supplies heating and cooling as the drying process requires. The process is considered more efficient as both the latent heat and sensible heat are recovered from the exhaust air improving the thermal efficiency.

- *Fluidized Bed Dryer* – The passing drying medium through the permeable support resting the feed material to be dried is utilized not just for drying but also for fluidization in this case. The superficial velocity of the drying medium, mostly heated air, is maintained such that the drying material is at the onset of fluidization and below the terminal velocity. This method is considered to be gentle, has a high degree of efficiency, and gives a uniform product. The thin boundary layer surrounding the particles to be dried due to rapid mixing ensures rapid heat and mass transfer, making it favourable for heat-sensitive materials. The rapid heat and mass transfer prevent the particles from getting overheated.

- *Heat Pump-Assisted Fluidized Bed Dryer* – These are similar to the fluidized bed systems but add a heat pump to control the humidity of the drying medium. These are still more efficient as it considers the advantage of rapid heat and mass transfer due to rapid mixing and thin layer over the particles to be dried and the improved thermal efficiencies due to heat recoveries.

Drying is a major operation in the agricultural and food industry. It is often used as a primary operation for the preservation of food materials or as a secondary process in some manufacturing operations. Drying is a higher energy-consuming process. Considering the complexity of the heat and mass transfer and associated physical and structural changes, it is ideal in such a scenario to have the flexibility for operating parameters and for handling different feed products (Delele et al., 2015). However, operational flexibility leads to complex designs and indirectly leads to higher capital can operating costs. Hence, the balance between technical feasibility and economic feasibility is crucial.

## 4.5 CONCLUSION

The fundamental principle of drying can be linked to nine impacting parameters. Several types of dryers can be designed and constructed to control these variable parameters and meet the critical drying requirement. In the case of dehydration of agricultural products, several dryer designs are already available. One should identify the specific requirements, understand the fundamentals & constraints and then proceed with selecting or modifying the dryer design for the specific aromatic crop. Since selective compounds in essential oils play a critical role for respective applications, extracting the same in primitive form is crucial. This also leads to criticalities of drying conditions so that the active ingredients are preserved in the natural matrix during the process and stored till extraction.

## REFERENCES

Barjinder, P. K., Nema, P. K., & Sharanagat, V. S. (2018). Fundamentals of drying. *Drying Technologies for Foods*, *28*, 1–22.

Chavan, A., Vitankar, V., Mujumdar, A., & Thorat, B. (2020). Natural convection and direct type (NCDT) solar dryers: A review. *Drying Technology*, *22*, 1–22. 10.1080/07373937. 2020.1753065

Chou, S. K., Chua, K. J., Mujumdar, A. S., Hawlader, M. N. A., & Ho, J. C. (2000). On the intermittent drying of an agricultural product. *Food and Bioproducts Processing: Transactions of the Institution of of Chemical Engineers, Part C*, *78*(4), 193–203. 10.1205/09603080051065296

Delele, M. A., Weigler, F., & Mellmann, J. (2015). Advances in the application of a rotary dryer for drying of agricultural products: A review. *Drying Technology*, *33*(5), 541–558. 10.1080/07373937.2014.958498

Gunathilake, D. D. M. C., Senanayaka, D. P., Adiletta, G., & Senadeera, W. (2018). Chapter 14 - Drying of Agricultural Crops. In G. Chen (Ed.), *Advances in Agricultural Machinery and Technology* (First, pp. 331–367). CRC Press Taylor & Francis Group.

Hayashi, H. (1989). Drying Technologies of Foods - Their History and Future. In *Drying Technology* (Vol. 7, Issue 2). 10.1080/07373938908916590

Mujumdar, A. S., & Devahastin, S. (2000). Chapter 1 - Fundamental Principles of Drying. In *Exergex* (1st ed., pp. 1–22).

Mujumdar, A. S., & Law, C. L. (2010). Drying technology: trends and applications in postharvest processing. *Food and Bioprocess Technology*, *3*(6), 843–852. 10.1007/s11 947-010-0353-1

Mujumdar, A. S. (2014). Section I - Fundamental Aspects. In A. S. Mujumdar (Ed.), *Handbook of Industrial Drying* (Forth, pp. 32–155). CRC Press Taylor & Francis Group.

Pai, K. R., Sindhuja, V., Ramachandran, P. A., & Thorat, B. N. (2021). Mass transfer "regime" approach to drying. *Industrial and Engineering Chemistry Research*, *60*(26), 9613–9623. 10.1021/acs.iecr.1c01680

# 5 Dehydration at Cellular Structures of Aromatic Plants

## 5.1 INTRODUCTION

Dehydration takes place through the application of heat and mass transfer at the cellular level. Though being explored and reported, a dearth of information is available on this topic, and it is not well understood. Drying is a multi- and interdisciplinary operation. As seen in the previous chapter, plant-based materials are complex & heterogeneous structures, porous, amorphous and hygroscopic by nature and have diverse chemical compositions and water-holding capacities. Such influential parameters sum up significant variables affecting simultaneous heat and mass transfer through the cellular structure.

Plant material microstructures are the geometric arrangements of cells, defined as spatial organizations at structural components and their interactions. Drying of plant material is basically the removal of water molecules (80%–90% content of plant tissues) from these micro-structures at three spaces of intercellular, intracellular and cell-wall levels. The quantification and ways of removing free-water, intercellular, and intracellular water need to be understood through experimental and mathematical analysis (Khan et al., 2017). The detailed analysis regarding the dehydration at the cellular level is not only valuable for determining the parameters affecting the drying, including activation energy, driving force/concentration gradient, internal & external mass transfer rate, and effective moisture diffusion (Majumder et al., 2021), but also it will help in the understanding of the conditions required for the retention of desired phytochemicals and volatiles in herbs, spices and aromatic plants (Pise et al., 2022; Prothon et al., 2003; Savitha et al., 2022).

Further, drying at the molecular level involves activating water molecules, mobilizing water molecules within the matrix, and transferring molecules outside the matrix. These three activities are carried out sequentially and at times simultaneously in dryers by controlling parameters such as the temperature of drying for the activation energy, mode of heating, internal diffusion and draft (natural or forced), and surface area exposure for external mass transfer (Khan et al., 2017; Mujumdar, 2014). The technical performances of the dryer, including drying rate, capacity, energy consumption and efficiency, as discussed in the previous chapter, that can be regulated, as required, through different dryer configurations and controlling individual parameters.

It is also highly desired to obtain the critical parameters for energy conservation vis-à-vis the desirable thermal conditions (Rahman et al., 2018). These parameters can be controlled by mode of energy application, temperature of the chamber, flow/ draft of & relative humidity of the drying medium and size of the material being dried. Necessary modifications in design can work out an efficient dehydration technique specific to a given application. Such alterations as per the region-specific conditions, local produces, and harvest seasons can significantly improve the eco-system, besides impactful socio-economic changeover (Orsat et al., 2008). Different drying techniques and the combination thereof have evolved, considering the products to be dried and meeting desired responses like colour, shrinkage, bulk density, porosity, phytochemical & volatile retention, antioxidants, sugars, proteins, and sensory attributes. Therefore, improving our understanding of these complex phenomena and developing useful prediction tools to study dehydration at the cellular level is important. Multistage and multi-mode dryers are outcomes of such systematic studies and research.

## 5.2   SIGNIFICANCE OF MICRO- AND MACRO-STRUCTURES

The physical properties of plant materials, such as leaves, fruit, and flowers, are important for controlling their metabolism, quality and the functioning of the plant. Plant materials maintain metabolic processes at the cellular level to retain their natural state and continue their functions. The cellular state of various plants is considered to be living. Cellular metabolism includes biochemical processes for energy creation, utilization and storage. These are mainly regulated by two organelles: the mitochondrion and the chloroplast. The death of a plant or cell can be considered when the micro-structure or plant material starts changing its form or starts degrading because of discontinued metabolism. On harvesting, the plant is known to continue with respiratory activity and preserve the integrity of the cellular microstructure. These structures impart mechanical properties and textures to the plant body. Over a certain period, typically the shelf life of the plant material, as the supply of water and nutrients is disrupted, the plant material starts degrading. For these reasons, the micro-structures play an important role in drying and extraction operations while processing medicinal and aromatic plants in the determined shelf life. The plant micro-structures are also responsible for supplying and removing water, gases and nutrients. This indicates an inherent mass transfer mechanism established at the cellular level. Hence, a detailed understanding of the plant micro- and macro-structures will not just facilitate the ideal pathways for mass transfer and heat transfer during the unit operations of dehydration, rehydration and extraction but will also help stabilize the microstructures to preserve the plant materials in the natural state.

As seen in the earlier chapter, macro-structures of the plant material may appear continuous, but at the microscopic level, the histological and cellular features like the type of tissues, the geometry of cells and tissues, presence of adhesive middle lamella, cellular water, cell wall structures, intercellular spaces and so on contribute individually as well as collectively to the properties. At a spatial scale, these features can be looked into at the *nanoscopic range* (plasmodesmata and plasma membrane), *microscopic range* (cell geometry, cell walls and lamella complex) and *macroscopic range* (plant parts and organs). To determine the diffusional properties (mass and thermal) of the plant materials, it is crucial to take into account the micro-scale geometry of tissue as well as intracellular spaces. These heterogeneous structures vary considerably in shape and size, with random connecting pathways, lengthy and complex at different scales in biological structures leading to complex geometries and spatial spaces for defining drying patterns. There is a lot of ongoing research on transport phenomenon at the microscopic level (Alabi et al., 2020), structural characterization (Liu and Cheng, 2011), modelling for volume changes (Purlis et al., 2021), multi-scale modelling (Mebatsion et al., 2008; Welsh et al., 2018) and so on; however, the relationship between the apparent macroscopic properties and the microscopic features can still be better defined. Consequently, the available continuum models have a limited range of validity, and the heterogeneity leads to extreme difficulties in understanding the drying and associated morphological changes during the process.

The product quality, typically the rehydration response, is closely related to these dehydration parameters, as seen at the micro-structural level (Savitha et al., 2022). The moisture removal rate must be well-balanced to control the quality of the dried product, typically the tissue integrity by minimizing the shear stresses during the process (Pise et al., 2022). The high-temperature drying and the more significant variation between the internal diffusion and evaporation rate can be catastrophic, causing permanent physical and chemical changes (Tanko et al., 2005). Herbs are leafy biomasses, having delicate structures, high moisture content, thin layers and high sensitivity towards temperature due to the presence of flavour, nutrients, colour, and texture of interest (Babu et al., 2018; Hossain et al., 2010). Hence, the understanding of the micro-structure of organelles, cells, tissues, and organs along with the role and functions (for *storage* of water/nutrient/gases/secreted metabolites or *transport channels* for the feed/waste/metabolites or *support structures* of the part) is required before proceeding with the dehydration or extraction unit processes.

## 5.2.1 Water at the Cellular Level in Plants

As seen in earlier chapters, the water held by different cellular environments, such as intercellular, intracellular and cell walls, differs according to their characteristics and operations (Welsh et al., 2018). The water residing in the plant material can be termed *capillary or free water* present in the intercellular spaces, *intracellular water* in the cells, and *cell wall water* in the fine spaces between the cell walls. The intracellular (loosely bound) and the cell wall water (strongly bound) are considered physically bound water. Plant materials are considered hygroscopic, amorphous and porous; hence, the bound water transport is considered to impact the plant material shrinkage during dehydration significantly (Khan et al., 2016; Prothon et al., 2003). The free water/intercellular water is suggested to have a low effect on the structural shrinkage on dehydration, loosely bound water is suggested to contribute to cellular shrinkage, pore formation and cell collapse and the strongly bound moisture migration results in tissue deformation. Considering these three, water-type migration, cellular shrinkage, tissue shrinkage, and structural collapse are mainly attributed (Joardder et al., 2015c, 2015a; Khan et al., 2017; Rahman et al., 2018).

Other than these three types of water, the chemical composition and physical structure of the plant material affect the heat and mass transfer involved during drying. The plant material can be classified as hygroscopic or non-hygroscopic based on the water binding capacity, porous or non-porous based on void-containing features, and amorphous or crystalline based on the organization of the matrix.

*Hygroscopic* materials are considered when a large amount of water is physically bound with the solid matrix. Most of the time, these materials tend to deform during the dehydration operation. These materials showcase an equilibrium moisture iso-thermal relation with vapour pressure different from pure water. The vapour pressure is dependent on the water activity of these hygroscopic materials. It can also be expressed as the water activity given at a certain moisture content or the degree of hygroscopicity (bond of water with solid matrix).

*Non-hygroscopic* material is not encompassed with bound water; hence, the vapour pressure is equal to the partial pressure of water. In such material, the pore spaces are completely filled with water on saturation and with air on complete dehydration. As the bound water in non-hygroscopic material is insignificant, the shrinkage seen during dehydration is negligible. In such materials, the vapour pressure depends only on temperatures, and the moisture movement leads to no further complications (Joardder et al., 2015b).

## 5.3   WATER DISTRIBUTION AND DRYING OF PLANT MATERIAL

The water in plant materials, mainly fresh fruits and vegetables, is predominantly seen in cells (intracellular water), and a small fraction is seen in the intercellular spaces. The mass transfer of these water molecules is seen as three mechanisms of transport of free water, transport of bound water, and transport of water vapours, and the flux can be considered intercellular flow, wall-to-wall flow and cell-to-cell flow. During dehydration, at the initial stage, as the moisture content is high, liquid flow is seen through the dominating capillary forces. Further, as the moisture content decreases, the gas phase builds up, lowering the liquid diffusion and changing over to vapour diffusion through porous structures. This water migration from the natural matrix, leaving voids behind, results in porosity (Joardder et al., 2015b). During drying operation, the rate of dehydration can be governed by either the internal mass diffusion or the external mass transfer. High temperature leads to faster water removal leading to unbalanced internal moisture diffusion and external mass transfer. However, due to the quicker mass transfer rate, the internal stresses at the cellular level are seen to increase, resulting in the loss of tissue integrity and hence volatiles (Pise et al., 2022; Savitha et al., 2022). The novel approach defined based on the drying by Pai et al. identifies the control over dehydration by knowing the relative rates of internal diffusion and external mass transfer rate (Chavan and Thorat, 2020; Pai et al., 2021).

During the dehydration of plant materials, various consideration like differences in the moisture content of the different cellular environment, the geometry of cells and tissues and the temperature of drying governs the morphological changes. Depending on the drying temperature, a number of primary properties, like the wet bulb/dry bulb temperature, diffusion rate, porosity, viscosity, and stiffness of the plant material structure. The temperature also leads to secondary consequences like drying efficiency, case hardening, morphological behaviours, pore stabilization, pore network development and so on. The temperature-independent parameters are the cell wall membrane porosity, cell fluid viscosity and cell wall contraction forces. For the ease of visualization and modelling of structures for the determination of drying patterns, the cell fluid (protoplasm) and cell walls (outer boundary as solid material) are taken into consideration. Tissues are considered an aggregation of these cells along with the lamella in the middle as intercellular spaces. So, during drying, this could be the simplest model for considering the changes in the moisture content, turgor pressure and cell wall contractions (Rathnayaka Mudiyanselage et al., 2017).

Morphological changes during drying are mainly represented by shrinkage and collapse of structures. The plant material matrix can be stated to be mainly of water, dry material and air. Volume change and large deformation occur in different biomasses during drying or water removal processing. Shrinkages, swelling and collapse, and deformation are well-known. Similarly, the breakdown of structures during post-drying processing also becomes vulnerable. Such phenomena result from complex and dynamic relationships between composition, microstructure, and driving forces established by operating conditions. Water plays a key role as a plasticizer, strongly influencing the state of amorphous materials via the glass transition and, thus, their mechanical properties.

An ideal dried product should reconstitute all properties of the original product on rehydration. For ideal drying conditions, the controlled activation energy for balanced internal mass diffusion and external mass transfer, low temperature and high surface areas are required, which may lead to a higher cost for the drying system. The most suitable dryer offers optimized conditions for higher capacity operation, better quality products, good efficiency, lower cost and minimal environmental impact (Khaing Hnin et al., 2019).

## 5.4 ROLE OF CELLULAR STRUCTURES/CELL WALL AND QUALITY OF DRIED PRODUCTS

Plant materials, as stated earlier, are complex and heterogeneous structures; hence, a fundamental understanding of the structures is required to predict, obtain or control the quality of the product during the drying or extraction unit operation. The associated heat and mass transfers during these processes depend on mechanical properties at various levels of plant structure. Plant materials include building blocks of proteins, glucose, water, air and minerals. The organelles, cells, tissues, and organs are complex combinations of these constituents in the form of fibres, polysaccharides, cellulose, hemicellulose, lignin, lipids and pectin. These building blocks help form the cell walls, cells and tissues in various shapes, with the intercellular spaces contributing to the porosity when filled with air (Figure 5.1).

These cells and tissues help contribute to the firm shape or texture of the structure. This firmness or structural integrity is due to the hydrostatic pressure, known as the turgor pressure, offered by the fluids in the cells. These fluids inside the cells (offering the turgor pressure), the fluid in the intercellular spaces and the

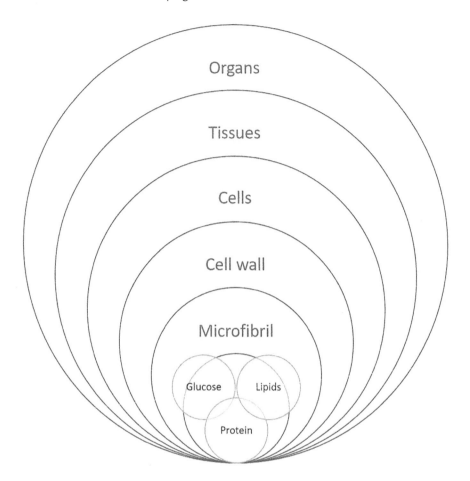

**FIGURE 5.1** Plant material structural features – molecular, cellular and tissue.

plasticizer fluids hold a strong relationship with the rheological properties of the plant material. Hence, water loss during dehydration directly impacts the micro-structures, spatial arrangement of cells, pores and tissues, resulting in structural shrinkage and collapse.

It is important to note that the plant material architecture and organization are complete and are considerate towards the required mass and heat transfer when living or growing. The evolution of plant materials is an efficient processing system. The cellular structure tends to alter and collapse only on harvesting or change in water or nutrient supply. While processing, the basic structural properties of porosity, shrinkage and diffusivity affect the quality of the product. The microstructures tend to change in two ways of shrinkage and collapse mainly. *Shrinkage* is defined as a reversible physical phenomenon in which a reduction in volume or size is seen due to the loss of intracellular water. In contrast, *collapse* is considered an irreversible breakdown at the cellular or tissue level. The volume shrinkage occurs during drying because of the negative pressure. The nature of shrinkage and collapse depends on the porosity of tissue intercellular adhesion and cell walls' strength. The deformation and the collapses are dependent on the processing conditions. These changes can be regulated by controlling the drying conditions.

The impact of drying temperatures on the structural integrity of secretory cells of betel leaves was observed and reported by Pise et al. (Pise et al., 2022). Dehydration at lower temperatures of 40°C and 50°C indicated that guard cells/mesophyll tissues protected the oil secretory cells (pearl glands) through controlled shrinkage. Low-temperature dehydration provided better control over cellular structures with lesser stress during the slower dehydration rate. At 60°C (a relatively higher temperature), faster dehydration resulted in similar shrinkage but with rupturing of the oil secretory cells (pearl glands). On rehydration the lack of structural retention confirmed the loss of integrity of the cells during dehydration and rehydration. It can be stated that a faster rate of dehydration leads to shrinkage and stiffening before the restructuring of the guard cells/mesophyll tissues. This leads to severe stresses on the adjacent delicate cells leading to the shearing of the cell walls and their rupturing. With the help of images, the significance of the slow rate of dehydration to protect the oil secretory cells (pearl glands) in aromatic leafy masses by the mesophyll tissues and cellular restructuring at the microscopic level are reported. Similar observations were reported by Savitha et al. on the microstructural dehydration of onion cells (Savitha et al., 2022).

The rheological properties of plant material impacting the shrinkage phenomenon during the drying operation are elastic modulus, hardness and stiffness. *Elastic modulus* is the measure of resistance to deformation when under stress. *Hardness* is defined as the resistance to plastic deformation, penetration, indentation or scratching. *Stiffness* is the quality of being firm, hard or unable to bend. These properties are interlinked, and the elastic modulus is related to the moisture content and the heterogeneity of the material (Karim et al., 2021). During dehydration, the variation in the elastic properties is observed as the moisture is reduced in the plant material.

Similarly, large shrinkages are initiated at cellular levels, bringing in another variable of the number of cell/tissue layers, shapes and size of the dehydrated plant material. This also leads to a detailed understanding of the shrinkage modelling and the multiscale studies of the drying operations. Several theoretical shrinkages models, like linear elastic models, elastoplastic models, visco-elastic models and hyper-elastic models, are being explored to understand the phenomenon of shrinkage and collapse during dehydration to achieve better control over the product quality (Mahiuddin et al., 2018). The plant structures, scale definitions, and multiscale modellings such as the multigrid model, equation-free framework and Heterogenous multiscale modelling are being explored (Welsh et al., 2018). Currently, the cellular properties of plant food material are being explored and simulated to develop an accurate model with some reported literature. However, the cellular properties of aromatic plant material are not reported much. Therefore, an extensive experimental investigation is needed to facilitate the development of an optimal dehydration and extraction protocol from aromatic plant material.

## 5.5 CONCLUSION

Water is an inevitable part of the plant material. Water, in the form of fluids, is observed in intercellular spaces and as plasticizer fluids or offering turgor pressure, providing firmness and structural integrity to the plant material. The process of releasing these fluids in the form of free and bound water and the drying rate are closely related to the morphological changes seen during the process. Water migration from the natural matrix, leaving voids behind, results in the porosity of the dried products. Post-dehydration, the stability of the pores, tissue intercellular adhesion and cell walls' strength, lead to either shrinkage or collapse of the plant material. The plant structure is inherently designed to transport feed/waste/metabolites. To efficiently dehydrate the plant biomass and retain the structural integrity for preserving desired secondary metabolites, a detailed understanding of the micro-structure of organelles, cells, tissues, and organs, along with their role and functions, is highly recommended before proceeding with the dehydration or extraction unit processes.

## REFERENCES

Alabi, K. P., Zhu, Z., Sun, D. W., 2020. Transport phenomena and their effect on microstructure of frozen fruits and vegetables. *Trends in Food Science and Technology* 101, 63–72. 10.1016/j.tifs.2020.04.016

Babu, A. K., Kumaresan, G., Raj, V. A. A., Velraj, R., 2018. Review of leaf drying: Mechanism and influencing parameters, drying methods, nutrient preservation, and mathematical models. *Renewable and Sustainable Energy Reviews* 90, 536–556. 10.1016/j.rser.2018. 04.002

Chavan, A., Thorat, B., 2020. Mathematical analysis of solar conduction dryer using reaction engineering approach. *International Journal of Chemical Reactor Engineering* 18. 10.1515/ijcre-2019-0220

Hossain, M. B., Barry-Ryan, C., Martin-Diana, A. B., Brunton, N. P., 2010. Effect of drying method on the antioxidant capacity of six Lamiaceae herbs. *Food Chemistry* 123, 85–91. 10.1016/j.foodchem.2010.04.003

Joardder, M. U. H., Brown, R. J., Kumar, C., Karim, M. A., 2015a. Effect of cell wall properties on porosity and shrinkage of dried apple. *International Journal of Food Properties* 18, 2327–2337. 10.1080/10942912.2014.980945

Joardder, M. U. H., Karim, A., Kumar, C., Brown, R. J., 2015b. Porosity: Establishing the relationship between drying parameters and dried food quality.

Joardder, M. U. H., Kumar, C., Brown, R. J., Karim, M. A., 2015c. A micro-level investigation of the solid displacement method for porosity determination of dried food. *Journal of Food Engineering* 166, 156–164. 10.1016/j.jfoodeng.2015.05.034

Karim, A., Fawzia, S., Rahman, M. M., 2021. Advanced micro-level experimental techniques for food drying and processing applications, *Advanced Micro-Level Experimental Techniques for Food Drying and Processing Applications.* 10.1201/9781003047018

Khaing Hnin, K., Zhang, M., Mujumdar, A. S., Zhu, Y., 2019. Emerging food drying technologies with energy-saving characteristics: A review. *Drying Technology* 37, 1465–1480. 10.1080/07373937.2018.1510417

Khan, M. I. H., Wellard, R., Pham, D. N., Karim, A., 2016. Investigation of cellular level of water in plant-based food material. *International Drying Symposium* 1, 7–13.

Khan, M. I. H., Wellard, R. M., Nagy, S. A., Joardder, M. U. H., Karim, M. A., 2017. Experimental investigation of bound and free water transport process during drying of hygroscopic food material. *International Journal of Thermal Sciences* 117, 266–273. 10.1016/j.ijthermalsci.2017.04.006

Liu, D., Cheng, F., 2011. Advances in research on structural characterization of agricultural products using atomic force microscopy. *Journal of the Science of Food and Agriculture* 91, 783–788. 10.1002/jsfa.4284

Mahiuddin, M., Khan, M. I. H., Kumar, C., Rahman, M. M., Karim, M. A., 2018. Shrinkage of Food materials during drying: Current status and challenges. *Comprehensive Reviews in Food Science and Food Safety* 17, 1113–1126. 10.1111/1541-4337.12375

Majumder, P., Sinha, A., Gupta, R., Sablani, S. S., 2021. Drying of selected major spices: Characteristics and influencing parameters, drying technologies, quality retention and energy saving, and mathematical models. *Food and Bioprocess Technology* 14, 1028–1054.

Mebatsion, H. K., Verboven, P., Ho, Q. T., Verlinden, B. E., Nicolaï, B. M., 2008. Modelling fruit (micro)structures, why and how? *Trends in Food Science and Technology* 19, 59–66. 1016/j.tifs.2007.10.003

Mujumdar, A. S., 2014. Section I - Fundamental aspects, in: Mujumdar, A. S. (Ed.), *Handbook of Industrial Drying*. CRC Press Taylor & Francis Group, pp. 32–155.

Orsat, V., Vijaya Raghavan, G. S., Sosle, V., 2008. Adapting drying technologies for agri-food market development in India. *Drying Technology* 26, 1355–1361. 10.1080/07373930802333452

Pai, K. R., Sindhuja, V., Ramachandran, P. A., Thorat, B. N., 2021. Mass transfer "regime" approach to drying. *Industrial and Engineering Chemistry Research* 60, 9613–9623. 10.1021/acs.iecr.1c01680

Pise, V., Shirkole, S., Thorat, B. N., 2022. Visualization of oil cells and preservation during drying of Betel Leaf (Piper betle) using hot-stage microscopy. *Drying Technology*. 10.1080/07373937.2022.2048848

Prothon, F., Ahrne, L., Sjoholm, I., 2003. Mechanisms and prevention of plant tissue collapse during dehdration - A Critical review. *Critical reviews in food sci* 43, 447–479.

Purlis, E., Cevoli, C., Fabbri, A., 2021. Modelling volume change and deformation in food products/processes: An overview. *Foods* 10, 778. 10.3390/foods10040778

Rahman, M. M., Kumar, C., Joardder, M. U. H., Karim, M. A., 2018. A micro-level transport model for plant-based food materials during drying. *Chemical Engineering Science* 187, 1–15. 10.1016/j.ces.2018.04.060

Rathnayaka Mudiyanselage, C. M., Karunasena, H. C. P., Gu, Y. T., Guan, L., Senadeera, W., 2017. Novel trends in numerical modelling of plant food tissues and their morphological changes during drying – A review. *Journal of Food Engineering* 194, 24–39. 10.1016/j.jfoodeng.2016.09.002

Savitha, S., Chakraborty, S., Thorat, B. N., 2022. Microstructural changes in blanched, dehydrated, and rehydrated onion. *Drying Technology* 40, 2550–2567. 10.1080/07373937.2022.2078347

Tanko, H., Carrier, D. J., Duan, L., Clausen, E., 2005. Pre- and post-harvest processing of medicinal plants. *Plant Genetic Resources* 3, 304–313. 10.1079/pgr200569

Welsh, Z., Simpson, M. J., Khan, M. I. H., Karim, M. A., 2018. Multi-scale Modeling for food drying: state of the art. *Comprehensive Reviews in Food Science and Food Safety* 17, 1293–1308. 10.1111/1541-4337.12380

# 6 Fundamental Principles and Technologies for Extraction

## 6.1 INTRODUCTION

Aromatic plants are recognized in the flavour and fragrance, pharmaceutical and FMCG industries owing to their broad spectrum of natural compounds and their wide range of therapeutic properties. The bioactive compounds present in plants are referred to as phytochemicals. These phytochemicals are derived from different parts of plants, such as leaves, bark, seed, seed coat, wood, flowers, fruit, roots and rhizomes. As described in chapter 2, each part has a different histology, function, and composition. Phytochemistry describes the occurrence of more than 600 secondary metabolic compounds in aromatic plants. Plants are the reservoirs of naturally occurring chemical compounds and structurally diverse bioactive molecules. The extraction of bioactive compounds from plants and their quantitative and qualitative estimation is important for the exploration of new biomolecules and applications in the industry directly or as a lead molecule to synthesize more potent molecules. Natural products like plant extract open a new horizon for discovering therapeutic agents. Extraction of these natural products from the plant is an empirical exercise since different solvents are utilized at varying conditions such as temperature, pressure, time and solvent flow. Bioactive components, marker compounds of key interest, extracted from the plants further need to be essentially separated from co-extractive compounds.

Typical techniques used for extraction from medicinal and aromatic plants include maceration, infusion, percolation, expression and effleurage (cold fat extraction), continuous hot extraction (Soxhlet), counter-current extraction, microwave-assisted extraction, ultrasound extraction (sonication), supercritical fluid extraction, and distillation techniques (water distillation, hydro-steam steam distillation. Some of the latest extraction methods for aromatic plants include headspace trapping, solid phase micro-extraction (SPME), protoplast extraction, and micro-distillation. Various factors have to be taken into consideration before proceeding with the selection of extraction methods, like part of the plant, degree of processing, moisture content, stability to heat, cost of extract, parameters for extraction (mainly time), the content of volatiles and final application. The solvent selection further depends on the extraction procedure, quality of phytochemicals to be extracted, diversity of desired and inhibitory compounds, ease of handling, toxicity and separation. (Azwanida, 2015; Ingle et al., 2017; Pandey et al., 2014).

DOI: 10.1201/9781003315384-6

## 6.2  METHODS IN EXTRACTION

Extraction of bioproducts means separating the complex mixtures of secondary metabolites from the secretory glands (special tissues), including osmophores, glandular trichomes ducts and cavities found on different parts in roughly 30% of the vascular plants (Sharifi-Rad et al., 2017). Therefore, the extraction process can be simplified as a series of steps on a laboratory or industrial scale. These include the *mass transfer* of the plant volatiles from the plant material to the solvent/carrier fluid, *separation* of the solvent/carrier fluid and the plant material, *separation* of the plant volatiles from the solvent/carrier fluid and *purification* of the essential/volatile oils.

The first unit operation of mass transfer can be carried out by different extraction methodologies using fluids like water/steam, organic solvents (polar or non-polar), critical fluids (sub- and super-), or fats and oils, or by mechanical means, with or without temperature, depending on the final applications. The complex mixture of compounds extracted, as plant volatiles (and non-volatiles in some cases), depends on various conditions at which the extraction is performed and partially on the extraction method. The extraction process can also be subjected to specific procedures and parameters for selective separation (Zhang et al., 2018). Appropriate extraction methods followed by necessary separation, isolation, and purification must be chosen to obtain volatile oils of commercial interest. Similarly, for getting essential oils, hydro-, steam- or dry-distillation, or mechanical process without heating (termed expression, especially for citrus oils) as per the definitions must be selected. The plant part's structure and density also affect the mass transfer kinetics and hence also need to be considered in case of selection of the extraction method.

Bioproducts, such as these volatiles, need assurance of constancy and quality to ensure safe and efficient operations for industrial applications, which is difficult to achieve. It is to be noted that the constituent of these volatiles varies with provinces, subspecies/chemotypes used, distillation period, stage of plant development, plant part used, and other factors (Figueiredo, 2017; Tisserand & Young, 2014). However, by limiting the variables like specific region, fixed maturity of harvests, same extraction process and uniform conditions, orderly pre-processing and so on, variation in the composition can be minimized. The combined effects of the constituents extracted lend to the oil characteristics such as odour and therapeutic properties. Hence, a thorough understanding of the extraction process, the impact of the extracts, and the reason for extraction/application are critical for determining the extraction process and obtaining the desired utilizable volatile oils.

### 6.2.1  DISTILLATION

Distillation is a method in which plant volatiles are separated using water or steam as extracting medium and then cooled to liquefy and separate hydrocarbons and water. The first part of separating volatiles from the plant material is rupturing the oil glands and releasing the oil using high-temperature steam/boiling water. Then, the vapour mixture acts as a carrier medium and is cooled for further oil and water separation based on density difference. The governing step in this process is the mass transfer of volatiles from plant material to steam, which occurs by physiochemical processes of *hydro-diffusion*. Depending on how the aromatic plant material is brought in contact with the water, the separation process is categorized into hydro-distillation, hydro-steam distillation or steam distillation (Baydar et al., 2008).

The simplest form of the distillation process, in which the plant material is boiled with water to accomplish hydro-diffusion, is as seen in Figure 6.1. Hydro-distillation is the oldest known process for essential oil extraction and is still widely used in many countries. Batch size is typical of 200 kg or 500 kgs plant material loaded with 3 to 4 times the amount of water. The still is directly heated by burning coal, wood, briquette, or electric heater. The hydro-steam distillation process is a slightly modified version of the hydro-distillation process. Here the extraction is carried out by steam as a carrying fluid but without direct contact with the boiling water and plant material. Steam is generated within the still, and the plant material is supported above the boiling water, as seen in Figure 6.1. Hydro-steam distillation is being implemented widely across the globe as the field distillation unit owing to the lower expenditure incurred for construction, operation and maintenance. It tends to give a better quality (lesser chances of charring) and slightly higher quantity of oil (hydrolysis of the ethers to a lesser extent). Hydrolysis reaction proceeds to an equilibrium point; hence, as less water is used in distillation, less is oil loss due to solubility and hydrolysis reaction. Steam distillation, sometimes termed *dry steam distillation,* is carried out by supporting the plant material in the distillation still and passing the steam generated in a satellite steam generator, as seen in Figure 6.1. The operating conditions of temperature and the pressure for the steam, being generated separately, can vary slightly, giving the process a versatile nature and key advantage. Steam distillation can be used

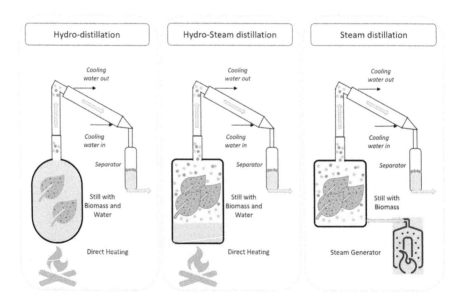

**FIGURE 6.1**    Schematic for essential oil extraction by hydro, hydro-steam and steam distillation.

for a wide range of aromatic plants, considering the flexibility in operating conditions. As the steam generation is remotely carried out, no direct heating source comes in contact with the plant material zeroing down the chances of thermal degradation or charring. Steam distillation is the most widely accepted process for producing essential oils on a large scale and is considered standard practice in the flavour and fragrance industries.

In all three cases, the essential oils are carried along with the steam through the top of the distiller into the condenser. The distillate is collected in copper stills similar to Florence flasks, and essential oils, lighter than water, are collected as a floating layer. The water is termed *hydrosols,* which can be used as perfumed water (Baydar et al., 2008).

In the above three processes, the first unit operation of volatile mass transfer from plant material, either by rupturing the glands or by hydro diffusion, occurs by steam/water and is the governing process. However, this mass transfer is a relatively slow process and consumes a high amount of energy. Drying plays a role in this case for reducing energy consumption for the distillation operation. The same is explained later in this chapter. Since the extraction of uncomminuted material takes a longer time than comminated material, there is a need for assisted methods for extraction, which have been developed recently to extract essential oils. The assisted extraction methods include ultrasound, pulsed electric field (high and moderate intensity), microwave-assisted extraction and DIC/auto vapourization.

## 6.2.2 ASSISTED EXTRACTION

Ultrasound (US) waves cause an implosion of cavitation bubbles in the medium, causing macro-turbulence, high-velocity inter-particle collisions and perturbation in micro-porous particles of the plant material. US waves help collapse the plant volatiles carrying tissues as well as causing accelerated eddy diffusion and internal diffusion resulting in the high mass transfer of the extractant (Vilkhu et al., 2008; Vinatoru, 2001). Biomass of aromatic plants has tissues in the form of glands containing the volatiles. During distillation, hydration and swelling followed by mass transfer of soluble compounds from material to extraction medium by diffusion were the key and rate-determining steps. In the presence of US waves, due to the implosions created by cavitation, the cell wall structure or the thin skin of the gland gets broken, releasing the oils and assisting the essential oil transfer in the extraction medium. US waves also enhance the extraction rate by facilitating the swelling and hydration of harder tissues. However, the medium should be below its boiling point to fulfil the primary purpose of gland breakdown. Hence, it only assists the extraction process and cannot be implemented as an independent extraction process (Assami et al., 2012; Balachandran et al., 2006; Vilkhu et al., 2008).

The use of electric fields initially began for the inactivation of microbes, later with the study of the increase in plant tissue permeability resulting in the extraction of cellular fluids. Application of high or moderate electric fields for milliseconds to microseconds can cause permeabilization of cell membranes. The charges on the cell membrane help form pores through which the fluids, in the case of aromatic, can extract the volatiles. The electric field is applied to the product by placing the electrodes with a moderate or high electric field ranging from 0.1 kV/cm to 50 kV/cm at room temperature (Chemat et al., 2015; Puértolas et al., 2011). The *Pulse Electric Field* (PEF) pre-treatment, a process for extracting active ingredients, gives the key advantage of the non-thermal performance, speediness, efficiency, low power and low pollution (Zhou et al., 2017). Furthermore, since there is minimal rupture of the tissue structures seen, there are minimum chances of having any other organic contaminants in the extract (Chemat et al., 2015). This technology is less explored for extracting essential oils pre-treatment for distillation. PEF has been reported as a pre-treatment method for extraction from rose (Zhou et al., 2017), nepeta, coriander seeds (Dobreva et al., 2013) and citronella grass (Hamzah et al., 2014). A noticeable increase in yield and significantly less time for extraction were reported in the studies (Figure 6.2).

Microwaves are electromagnetic radiations with a frequency from 0.3 GHz to 300 GHz possessing electric and magnetic fields perpendicular to each other. The heating occurs by an electric field through dipolar rotation and ionic conduction. The alignment on the electric field of the molecules, both of the solvent and the sample possessing a dipole moment, causes the dipolar rotation. The high-frequency oscillations cause the collision of the molecules liberating the heat in the medium resulting in very fast heating. The larger the dielectric constant more optimal the heating is seen. The effect of or the utilization of microwave energy is strongly dependent on the solvent, and the solid matrix and the solvents are selected based on their superior dielectric constant (Christen & Kaufmann, 2002). Microwave extraction can be implemented for fresh aromatic plants or plant material as they have enough moisture content for a solvent-free extraction

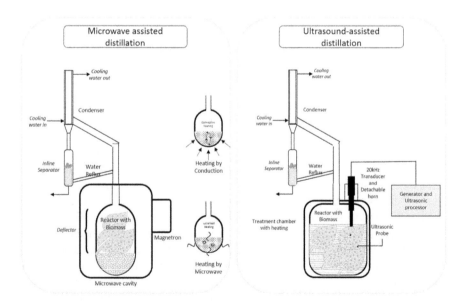

**FIGURE 6.2** Schematic for assisted more of extraction for essential oils – microwave assisted & ultrasound assisted extraction.

technique. When exposed to microwave radiation which heats the in-situ water, it distends, leading to the rupture of glands freeing the volatiles. The same water, converted to steam, now helps carry the volatiles, cooled and separated to get the desired product (Lucchesi et al., 2004). This principle of microwave heating is used in multiple ways like microwave-assisted solvent extraction (MASE) (Cardoso-Ugarte et al., 2013), vacuum microwave hydro-distillation (VMHD) (Bustamante et al., 2016), microwave hydro-distillation (MWHD) (Stashenko et al., 2004), compressed air microwave distillation (CAMD) (Craveiro et al., 1989), solvent-free microwave hydro-distillation (SFME) (Lucchesi et al., 2004) and Microwave hydro diffusion and gravity (MHG) (Vian et al., 2008).

*Détente Instantané Contrôlée* (DIC) is mainly based on the principle of thermodynamics of instantaneity. In this high-temperature, short-time process, the fast transition from high-pressure steam to vacuum causes thermos-mechanical stresses resulting in the release of the components for the plant matrix (Chemat et al., 2015). The two main steps involved in this process are heating using high-pressure steam (up to 1 MPa and 180°C) and then abrupt pressure drop towards the vacuum at a depressurization rate higher than 0.5 kPa/s in a short time of about 5–60 s (Naji et al., 2008).

### 6.2.3 SOLVENT EXTRACTION

Solvent extraction is a term used to describe a process of solute mass transfer from the solid/liquid phase to the solvent phase occurring based on affinity towards the solvent. The solvent penetrates plant material and dissolves natural aromatic compounds with some waxes, albumins and colouring matter, which gives concretes, resinoids, pomades and absolutes on further solvent separation. Absolutes are alcohol-soluble compounds obtained from concretes, post-separation of waxes, having a similar odour to essential oils. This process is widely employed to extract aromatic components from flowers, leaves, and mosses.

The key processes involved in solvent extraction are solute diffusion and the change of solute phase; the extraction process is mainly influenced by particle size, solvent type and polarity, temperature, and agitation. The solvent extraction process can be carried out as batch and continuous processes (Prado et al., 2015). Industrially, the solvent extraction process is executed in two ways: stationary and rotary. In the *stationary system*, aromatic plant material is loaded onto the grids in a vertical cylinder and given two to three washes of fresh solvents to extract the perfume-rich solvents. In the *rotary system*, a horizontally rotating tinned drum is used for charging the plant material and rinsing it with solvent by turning. Concrete obtained from solvent evaporation is washed with alcohol to separate waxes (Singh, 2008) (Figure 6.3).

Solvent selection in this process for extraction plays a critical role. The choice of solvent depends on the characteristics like favourable distribution coefficient for a product, selectivity, low emulsion forming tendency, low aqueous solubility, chemical and thermal stability, product and solvent recoverability and cost and bulk availability (Bruce & Daugulis, 1991; Singh, 2008) (Table 6.1).

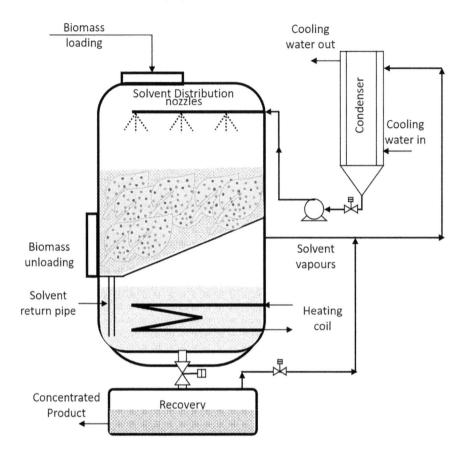

**FIGURE 6.3**    Schematic for solvent extraction of volatile oils from aromatic plants material.

### 6.2.4 PRESSURIZED/ACCELERATED SOLVENT AND SUBCRITICAL FLUID EXTRACTION

The solvent extraction process is further enhanced by having pressurized/acceler-
ated solvent extraction (PSE/ASE) and subcritical fluid extractions. PSE/ASE uses
organic solvents at elevated temperatures (50°C to 200°C) to enhance the extraction
kinetics and high pressure (10–15 MPa) to keep solvent in the liquid phase, thus
achieving a rapid and safe extraction (Tandon & Rane, 2008). Increased temperature accelerates the extraction kinetics, and elevated pressure keeps the solvent in
the liquid state, thus enabling safe and rapid extractions.

Subcritical water extraction is a powerful new technique with temperatures
between 100°C and 374°C and high pressure to maintain the liquid state. Water has
unique properties of a high boiling point for its mass, a high dielectric constant and
high polarity. With high temperatures, there is a decrease in permittivity, an
increase in diffusivity and a decrease in the viscosity and surface tension, with the
help of which moderately polar and non-polar compounds can be extracted. Water
is a highly polar solvent and has a high dielectric constant of 80 at 25°C; however,
in the subcritical range, the hydrogen bonds are broken, and the dielectric constant

**TABLE 6.1**

**Solvents for Extraction of Phytochemicals and the Threshold Limits in the Extract**

| Solvent | Boiling Point, °C | Miscibility *with H2O* | Threshold Limit *Values, ppm* |
|---|---|---|---|
| Acetone | 56 | Completely miscible | 1000 |
| Acetic acid | 116–117 | Completely miscible | 10 |
| Ethyl acetate | 77 | 80% | 400 |
| Benzene | 80 | <0.01% | 25 |
| 2-Butanol | 79.5 | 19% | 2200 |
| Cyclohexane | 80.7 | <0.01% | 300 |
| Dichloromethane | 39.7 | 1.30% | 2200 |
| Chloroform | 61 | 8% | 50 |
| Carbon tetrachloride | 76.77 | 0.80% | 10 |
| Hexane | 69 | <0.01% | – |
| Ethanol | 78 | Completely miscible | 1000 |
| Ethyl ether | 34.6 | 1.20% | 400 |
| Petrol ether | 30–50 | – | 500 |
| Propanetriole | 290* | Completely miscible | – |
| Methanol | 64.7 | Completely miscible | 200 |
| 1-Propanol | 91 | Miscible | 400 |
| 2-Propanol | 82.4 | Miscible | 400 |
| Toluene | 110.6 | 0.06 | 100 |

is reduced to 25 at 250°C and 25 bar pressure, helping in dissolving medium and low polar compounds (Haghighi & Khajenoori, 2013).

Supercritical fluid extraction (SFE) is the next phase of pressurized solvent extraction. With the increase in temperature, the liquid density diminishes, and an increase in pressure increases the vapour density converging at the critical point where there is no distinction between vapour and liquid. SFE offers a new dimension with very low surface tensions, low viscosities and moderately high diffusion coefficients with better solvent power for extraction (Pereda et al., 2008) (Figures 6.4 and 6.5) (Table 6.2).

Supercritical fluids, due to the ecological benefits of low energy consumption, are considered green solvents for the future. Supercritical carbon dioxide and water are the most preferred solvents considering their non-carcinogenic, non-toxic, non-mutagenic, non-flammable and thermodynamically stable properties (Marr & Gamse, 2000). In addition, the key advantage of the ease of separation makes it acceptable to obtain natural extracts (Knez et al., 2019; Reverchon & Marco, 2008). SFE extraction is similar to solvent extraction but is mainly carried out by modification of the thermodynamic properties. The plant material is loaded in an extractor, and the extracted components are separated from the solvent in the separator. The pump and the valve system is considered to alter the thermodynamic properties of the extractor and separator. Extractor generally operates at a higher

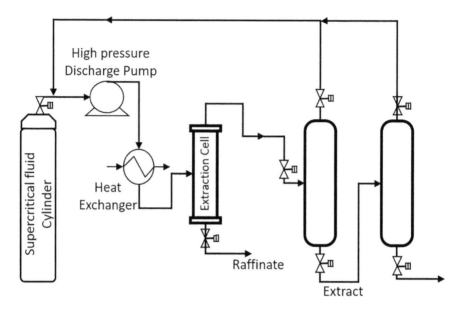

**FIGURE 6.4**    Schematic for supercritical fluid extraction of volatile oils from aromatic plant materials.

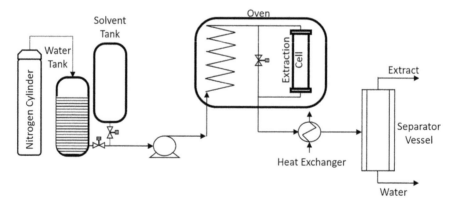

**FIGURE 6.5**    Schematic for pressurized solvent/subcritical fluid extraction of volatile oils from aromatic plant materials.

pressure of (100–600 bar) and desired extraction temperature. The pressurized high-temperature solvent is depressurized sequentially in single or multiple separators operated at different values (intermediate value @ 150–200 bar and low pressure @ 30–50 bar) using pressure control valves to obtain the extracted oleoresins as a complete mix or with component separation (Viplav H Pise & Thorat, 2022; Reverchon & Marco, 2008).

The parameters affecting the extraction conditions and the impacts are listed on the following page (Table 6.3).

## TABLE 6.2
## Fluids Considered for Supercritical Fluid Extraction and their Critical Operating Conditions

| Fluid | Critical Temperature $T_c/K$ | Critical Pressure $P_c/bar$ | Critical Volume $V_c/cm^3 \cdot mol^{-1}$ |
|---|---|---|---|
| $CO_2$ | 304.12 | 73.7 | 94.07 |
| Ethane | 305.3 | 48.7 | 145.5 |
| Propane | 369.8 | 42.5 | 200 |
| Ammonia | 405.4 | 113.5 | 72.47 |
| n-Hexane | 507.5 | 30.2 | 368 |
| Methanol | 512.6 | 80.9 | 118 |
| Water | 647.1 | 220.6 | 55.95 |

## TABLE 6.3
## Influencing Parameters for Solvent Extraction and their Potential Impacts

| Parameters | Impact |
|---|---|
| Pressure | Solubility of the components increases with an increase in pressure. |
| Temperature | • Solubility of the compound increases with a decrease in temperature close to critical pressure.<br>• Solubility of the compound increases with an increase in temperature at higher pressure. |
| Time | The higher the time, the better the possibility of meeting the saturation concentration of the solvent at operating conditions. |
| Solvent flowrate (1mm/s to 5 mm/s) | Optimization is required based on consideration for impacts as below:<br>• A higher flowrate implies higher operating and capital costs.<br>• Lower flowrates imply lower production capacity.<br>• Higher flowrate results in lower residence time, causing lower loading of solute in the solvent than the saturation concentration at operating conditions. |
| Solvent-to-feed ratio (30:1 to 100:1) | To be decided on factors like:<br>• Concentration in the feed material<br>• Solubility of the components in a supercritical solvent<br>• Type of feed<br>• Distribution of the components in the feed |

### 6.2.5 MOLECULAR DISTILLATION

Molecular distillation (MD), also known as short-path distillation, is a well-established technique. MD is a non-equilibrium process in which the evaporating surface of the still is very close to the condensing surface, giving a very short distance for the molecule to travel on evaporation and under very low-pressure conditions, the distance travelled is comparable to the mean free path of the molecules. Hence, this distillation process is called *molecular distillation*. MD is a post-extraction process used for the purification and separation of natural extracts and for efficiently separating absolutes and waxes. The other processes carried out using MD are deodorizing oils and separating fragrances derived from fatty acids (Pangarkar, 2008).

The MD process is sensitive towards, and its efficiency is crucially affected by the amount of low-boiling volatiles and presence of air, moisture, or other gasses in the feed, the temperature difference between the condensing and evaporating surfaces, viscosity of the feed and operating pressures. Hence, considering the sensitivity of the process, care should be taken to meet the prerequisites of the process (Pangarkar, 2008).

Out of the processes mentioned above, it can be said that steam distillation, solvent extraction, supercritical fluid extraction and molecular distillation (for post-processing) are mature processes. In contrast, subcritical water extraction, pressurized/accelerated solvent extraction, microwave extraction, and ultrasound extraction methods are in the growth phase. Pulsed electric field extraction and auto-vaporization are in the research phase. In addition, there are other extraction processes, like micellar extraction (Malysa-Pasko et al., 2018) and enzyme extraction (pre-treatment methods) (Hosni et al., 2013), which are also in the research phase.

Various extraction methods have been discussed, targeting different extraction steps by different principles to achieve better efficiency and cost-effective processing of aromatic plants. The selection and implementation of processes on an industrial scale are made stepwise (considering product quality requirements, extraction yield anticipated, processing capacity and so on). Comparison based on capital investment, operating costs and product pricing (based on quality) leads to final implementation based on Cost & Revenue estimation. Presently, with the increasing environmental concerns, customers' interest in natural products, transparency and traceability of the product from source, and fair-trade practices, the selection of process is governed by these factors besides the Cost & Revenue estimation (Burger et al., 2019). A schematic for farm-based distillation, collection, purification and isolation is shown in Figure 6.6.

## 6.3 ROLE OF DRYING IN EXTRACTION

As seen in earlier chapters, the moisture content in aromatic plants may range from 30% to 95%. One of the facts to be considered while dehydration is how to retain the structural integrity and provide a pathway for extracting the secondary metabolites from the plant matrix. Essential oils, by definition, are "products obtained from natural raw materials of vegetal origin, by steam distillation, or dry distillation, after the separation of the aqueous phase – if any – by physical

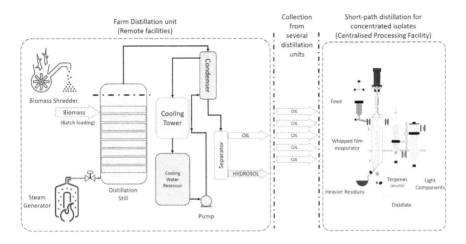

**FIGURE 6.6** Schematic for steam distillation unit and central processing by short path distillation for purification and isolation.

process." Considering the definition, steam is a non-negotiable part of the extraction process. Steam generation, similar to drying, is an energy-intensive process. This makes essential oil extraction a volume as well as energy intensive process (Viplav Hari Pise & Thorat, 2023). However, in the distillation process, drying can support reducing the steam load and inclining it towards the sustainable process.

The steam distillation process can occur in several steps, including increasing the temperature of the plant material, rupturing the secretary glands, transferring the metabolites from the plant matrix to the carrying fluid and carrying the metabolites from the distillation still to the separator. Steam, considered for extraction, is performing the role of providing the heat duty to increase the temperature, rupturing the tissues and carrying the volatiles. The heat duty required for increasing the temperature itself is significant as it consists of increasing the temperature of the associated equipment and the plant material. Fresh biomass, if considered, accounts for a significant quantity of water that needs to be heated and vaporized. On controlled dehydration of these aromatic plant materials, the moisture content in the plant can significantly be reduced, which leads to a lowering of the heat duty requirement for increasing the temperature and vaporization of the water.

Table 6.4 showcases a case for dehydrating geranium leaves and subjecting them to steam distillation. The total steam requirement is calculated for increasing the temperature of equipment and plant material at different moisture contents and the steam required as the carrying medium. It can be seen that, based on the moisture content of the plant material, a steam saving in the range of 50%–86% can be achieved for the plant material heating, which leads to an overall steam saving of up to 23.92% per batch of 1,000 kg plant material subject to extraction. This also confirms that dehydration plays a critical role in energy saving during the extraction of essential oils.

**TABLE 6.4**

**Steam Saving on Dehydration of Plant Material (Geranium) During Steam Distillation**

| Description | Units | Cases | | | | | |
|---|---|---|---|---|---|---|---|
| Fresh plant material (MC of 84%) | kg | 1000 | 1000 | 1000 | 1000 | 1000 | 1000 |
| Dry plant material | kg | 160 | 160 | 160 | 160 | 160 | 160 |
| Moisture content post-controlled drying | % | 84% | 68% | 60% | 50% | 32% | 20% |
| The initial weight of a batch | kg | 1000 | 500 | 400 | 320 | 235.29 | 200 |
| Enthalpy of water | J/kg.K | 4186.00 | 4186.00 | 4186.00 | 4186.00 | 4186.00 | 4186.00 |
| Enthalpy of dry plant material | J/kg.K | 2263.00 | 2263.00 | 2263.00 | 2263.00 | 2263.00 | 2263.00 |
| Enthalpy of plant material | J/kg.K | 3878.32 | 3570.64 | 3416.80 | 3224.50 | 2878.36 | 2647.60 |
| Initial temperature | Deg C | 25.00 | 25.00 | 25.00 | 25.00 | 25.00 | 25.00 |
| Final temperature | Deg C | 110.00 | 110.00 | 110.00 | 110.00 | 110.00 | 110.00 |
| Steam pressure | barg | 1.5 | 1.5 | 1.5 | 1.5 | 1.5 | 1.5 |
| Specific enthalpy of superheated steam @ 1.5 barg | kJ/kg | 2716.23 | 2716.23 | 2716.23 | 2716.23 | 2716.23 | 2716.23 |
| Specific enthalpy of water @ 0.5 barg | kJ/kg | 467.73 | 467.73 | 467.73 | 467.73 | 467.73 | 467.73 |
| Utilisable enthalpy up to 0.5 barg Sat. Liq. | kJ/kg | 2248.5 | 2248.5 | 2248.5 | 2248.5 | 2248.5 | 2248.5 |
| Enthalpy requirement for plant material heating | kJ | 3296557.2 | 151752.2 | 116171.2 | 87706.4 | 57567.2 | 45009.2 |
| The overall efficiency of the process | % | 50% | 50% | 50% | 50% | 50% | 50% |
| Steam requirement | kg | 293.22 | 134.98 | 103.33 | 78.01 | 51.20 | 40.03 |
| Steam saving for plant material heating | | 0.00% | 53.97% | 64.76% | 73.39% | 82.54% | 86.35% |
| Metal | kg | 250.00 | 250.00 | 250.00 | 250.00 | 250.00 | 250.00 |
| Initial temp | °C | 25.00 | 25.00 | 25.00 | 25.00 | 25.00 | 25.00 |
| Final temp | °C | 110.00 | 110.00 | 110.00 | 110.00 | 110.00 | 110.00 |

| | | | | | | | |
|---|---|---|---|---|---|---|---|
| $C_p$ | J/kg. K | 490.00 | 490.00 | 490.00 | 490.00 | 490.00 | 490.00 |
| Enthalpy required for unit heating | kJ | 10412.50 | 10412.50 | 10412.50 | 10412.50 | 10412.50 | 10412.50 |
| The overall efficiency of the process | % | 75% | 75% | 75% | 75% | 75% | 75% |
| Steam requirement | kg | 6.17 | 6.17 | 6.17 | 6.17 | 6.17 | 6.17 |
| | | | | | | | |
| Essential oil content | % | 0.15% | 0.15% | 0.15% | 0.15% | 0.15% | 0.15% |
| Essential oil extracted | kg | 1.5 | 1.5 | 1.5 | 1.5 | 1.5 | 1.5 |
| The density of geranium essential oil | g/ml | 0.889 | 0.889 | 0.889 | 0.889 | 0.889 | 0.889 |
| Steam requirement for carrying essential oils | g/ml | 450 | 450 | 450 | 450 | 450 | 450 |
| Steam required for carrying essential oil per batch | kg | 759.28 | 759.28 | 759.28 | 759.28 | 759.28 | 759.28 |
| | | | | | | | |
| Total steam required per batch | kg | 1058.68 | 900.44 | 868.79 | 843.47 | 816.66 | 805.49 |
| *Steam saving per batch* | | *0.00%* | *14.95%* | *17.94%* | *20.33%* | *22.86%* | *23.92%* |

## 6.4 CONCLUSION

The complexity of the plant-based aroma and essential oils, considering the chemical composition, is well recognized. Since selective compound in the oil plays a critical role in the respective application, extraction of the same in primitive form is crucial. The selectivity of the compounds and extraction efficiency can be controlled and optimized by considering several parameters in extraction, like solvent polarity, temperature and pressure of extraction, solvent flow rates, time for extraction, and so on. A detailed understanding of the extraction process is required to control each of these parameters systematically and optimize an efficient extraction.

Similarly, dehydration of plant biomass helps lower the heat duty requirement of the overall extraction process inclining it more towards a sustainable process.

## REFERENCES

Assami, K., Pingret, D., Chemat, S., Meklati, B. Y., & Chemat, F. (2012). Ultrasound induced intensification and selective extraction of essential oil from Carum carvi L. seeds. *Chemical Engineering and Processing: Process Intensification*, *62*, 99–105. 10.1016/j.cep.2012.09.003

Azwanida, N. N. (2015). A Review on the extraction methods use in medicinal plants, principle, strength and limitation. *Medicinal & Aromatic Plants*, *04*(03), 3–8. 10.4172/2167-0412.1000196

Balachandran, S., Kentish, S. E., Mawson, R., & Ashokkumar, M. (2006). Ultrasonic enhancement of the supercritical extraction from ginger. *Ultrasonics Sonochemistry*, *13*(6), 471–479. 10.1016/j.ultsonch.2005.11.006

Baydar, H., Schulz, H., Krüger, H., Erbas, S., & Kineci, S. (2008). Influences of fermentation time, hydro-distillation time and fractions on essential oil composition of damask rose (rosa damascena mill.). *Journal of Essential Oil-Bearing Plants*, *11*(3), 224–232. 10.1080/0972060X.2008.10643624

Bruce, L. J., & Daugulis, A. J. (1991). Solvent Selection strategies for extractive biocatalysis. *Biotechnology Progress*, *7*(2), 116–124. 10.1021/bp00008a006

Burger, P., Plainfossé, H., Brochet, X., Chemat, F., & Fernandez, X. (2019). Extraction of natural fragrance ingredients: History overview and future trends. *Chemistry and Biodiversity*, *16*(10). 10.1002/cbdv.201900424

Bustamante, J., van Stempvoort, S., García-Gallarreta, M., Houghton, J. A., Briers, H. K., Budarin, V. L., Matharu, A. S., & Clark, J. H. (2016). Microwave assisted hydro-distillation of essential oils from wet citrus peel waste. *Journal of Cleaner Production*, *137*, 598–605. 10.1016/j.jclepro.2016.07.108

Cardoso-Ugarte, G. A., Juárez-Becerra, G. P., Sosa-Morales, M. E., & López-Malo, A. (2013). Microwave-assisted extraction of essential oils from herbs. *Journal of Microwave Power and Electromagnetic Energy*, *47*(1), 63–72. 10.1080/08327823.2013.11689846

Chemat, F., Fabiano-Tixier, A. S., Vian, M. A., Allaf, T., & Vorobiev, E. (2015). Solvent-free extraction of food and natural products. *TrAC - Trends in Analytical Chemistry*, *71*, 157–168. 10.1016/j.trac.2015.02.021

Christen, P., & Kaufmann, B. (2002). Recent Extraction techniques for natural products: Microwave-assisted extraction and pressurised solvent extraction. *Phytochemical Analysis*, *113*(Phytochem. Anal. 13), 105–113. University of Geneva, School of Pharmacy, Laboratory of Pharmaceutical Analytical Chemistry, 20 bd d'Yvoy, CH-1211 Geneva 4, Switzerland

Craveiro, A. A., Matos, F. J. A., Alencar, J. W., & Plumel, M. M. (1989). Microwave oven extraction of an essential oil. *Flavour and Fragrance Journal*, *4*(1), 43–44. 10.1002/ffj.2730040110

Dobreva, A., Tintchev, F., & Toepfl, S. (2013). Effect of pulsed electric fields on distillation of essential oil crops. *CHIMIE Biotechnology*, *9*(January), 1255–1260.

Figueiredo, A. C. (2017). Biological properties of essential oils and volatiles: Sources of variability. *Natural Volatiles and Essential Oils*, *4*(4), 1–13.

Haghighi, A., & Khajenoori, M. (2013). Subcritical Water Extraction. In *Mass Transfer - Advances in Sustainable Energy and Environment Oriented Numerical Modeling* (pp. 459–488). 10.5772/54993

Hamzah, M. H., Che Man, H., Abidin, Z. Z., & Jamaludin, H. (2014). Comparison of citronella oil extraction methods from Cymbopogon nardus grass by ohmic-heated hydro-distillation, hydro-distillation, and steam distillation. *BioResources*, *9*(1), 256–272. 10.15376/biores.9.1.256-272

Hosni, K., Hassen, I., Chaâbane, H., Jemli, M., Dallali, S., Sebei, H., & Casabianca, H. (2013). Enzyme-assisted extraction of essential oils from thyme (Thymus capitatus L.) and rosemary (Rosmarinus officinalis L.): Impact on yield, chemical composition and antimicrobial activity. *Industrial Crops and Products*, *47*(2013), 291–299. 10.1016/j.indcrop.2013.03.023

Ingle, K. P., Deshmukh, A. G., Padole, D. A., Dudhare, M. S., Moharil, M. P., & Khelurkar, V. C. (2017). Phytochemicals: Extraction methods, identification and detection of bioactive compounds from plant extracts. *Journal of Pharmacognosy and Phytochemistry*, *6*(1), 32–36. https://www.phytojournal.com/archives/2017.v6.i1.1058/phytochemicals-extraction-methods-identification-and-detection-of-bioactive-compounds-from-plant-extracts

Knez, Ž., Pantić, M., Cör, D., Novak, Z., & Knez Hrnčič, M. (2019). Are supercritical fluids solvents for the future? *Chemical Engineering and Processing - Process Intensification*, *141*(December 2018), 1–8. 10.1016/j.cep.2019.107532

Lucchesi, M. E., Chemat, F., & Smadja, J. (2004). Solvent-free microwave extraction of essential oil from aromatic herbs: Comparison with conventional hydro-distillation. *Journal of Chromatography A*, *1043*(2), 323–327. 10.1016/j.chroma.2004.05.083

Malysa-Pasko, M., Lukasiewicz, M., Ziec, G., Slusarz, E., & Jakubowski, P. (2018). Micellar extraction of active ingredients of plant raw materials as a tool for improving the quality of diet supplements and additional substances. *Proceedings*, *9*(1), 59. 10.3390/ecsoc-22-05789

Marr, R., & Gamse, T. (2000). Use of supercritical fluids for different processes including new developments - a review. *Chemical Engineering and Processing: Process Intensification*, *39*(1), 19–28. 10.1016/S0255-2701(99)00070-7

Naji, G., Mellouk, H., Rezzouget, S. A., & Allaf, K. (2008). Extraction of essential oils of juniper berries by instantaneous controlled pressure-drop: Improvement of dic process and comparison with the steam distillation. *Journal of Essential Oil-Bearing Plants*, *11*(4), 356–364. 10.1080/0972060X.2008.10643641

Pandey, A., Tripathi, S., & Pandey, C. A. (2014). Concept of standardization, extraction and pre phytochemical screening strategies for herbal drug. *Journal of Pharmacognosy and Phytochemistry JPP*, *115*(25), 115–119.

Pangarkar, V. G. (2008). Microdistillation, Thermomicrodistillation and Molecular Distillation Techniques. In S. S. Handa, S. P. S. Khanuja, G. Longo, & R. Dev Dutt (Eds.), *Extraction Technologies for Medical and Aromatic Plants* (pp. 129–143). International Centre For Science And High Technology.

Pereda, S., Bottini, S. B., & Brignole, E. A. (2008). Fundamentals of Supercritical Fluid Technology. In J. L. Martínez (Ed.), *Supercritical Fluid Extraction of Nutraceuticals and Bioactive Compounds* (pp. 1–25). CRC Press.

Pise, V. H., & Thorat, B. N. (2022). Supercritical fluid extraction of dried Surangi flowers (Mammea suriga). *Industrial Crops & Products*, *186*(April), 115268. 10.1016/j.indcrop.2022.115268

Pise, V. H., & Thorat, B. N. (2023). Green steam for sustainable extraction of essential oils using solar steam generator: A techno-economic approach. *Energy Nexus*, *9*(January), 100175. 10.1016/j.nexus.2023.100175

Prado, J. M., Vardanega, R., Debien, I. C. N., Meireles, M. A. de A., Gerschenson, L. N., Sowbhagya, H. B., & Chemat, S. (2015). Chapter 6 – Conventional Extraction. In *Food Waste Recovery* (pp. 127–148). Elsevier Inc. 10.1016/B978-0-12-800351-0.00006-7

Puértolas, E., Saldaña, G., Álvarez, I., & Raso, J. (2011). Experimental design approach for the evaluation of anthocyanin content of rosé wines obtained by pulsed electric fields. Influence of temperature and time of maceration. *Food Chemistry*, *126*(3), 1482–1487. 10.1016/j.foodchem.2010.11.164

Reverchon, E., & Marco, I. De. (2008). Essential Oils Extraction and Fractionation Using Supercritical Fluids. In J. L. Martínez (Ed.), *Supercritical Fluid Extraction of Nutraceuticals and Bioactive Compounds* (pp. 305–337). CRC Press.

Sharifi-Rad, J., Sureda, A., Tenore, G. C., Daglia, M., Sharifi-Rad, M., Valussi, M., Tundis, R., Sharifi-Rad, M., Loizzo, M. R., Oluwaseun Ademiluyi, A., Sharifi-Rad, R., Ayatollahi, S. A., & Iriti, M. (2017). Biological Activities of Essential Oils: From Plant Chemoecology to Traditional Healing Systems. In *Molecules* (Vol. 22, Issue 1). 10.3390/molecules22010070

Singh, J. (2008). Maceration, Percolation and Infusion Techniques for the Extraction of Medicinal and Aromatic Plants. In S. S. Handa, S. P. S. Khanuja, G. Longo, & R. Dev Dutt (Eds.), *Extraction Technologies for Medicinal and Aromatic Plants* (Vol. 1, pp. 67–82). International Centre For Science And High Technology. 10.1038/ja.2011.74

Stashenko, E. E., Jaramillo, B. E., & Martínez, J. R. (2004). Analysis of volatile secondary metabolites from Colombian Xylopia aromatica (Lamarck) by different extraction and headspace methods and gas chromatography. *Journal of Chromatography A*, *1025*(1), 105–113. 10.1016/j.chroma.2003.10.059

Tandon, S., & Rane, S. (2008). Decoction and Hot Continuous Extraction Techniques. In S. S. Handa, S. P. S. Khanuja, G. Longo, & R. Dev Dutt (Eds.), *Extraction Technologies for Medical and Aromatic Plants* (pp. 93–106). International Centre For Science And High Technology.

Tisserand, R., & Young, R. (2014). Essential Oil Composition. In *Essential Oil Safety* (2nd ed.). © 2014 Robert Tisserand and Rodney Young. 10.1016/b978-0-443-06241-4.00002-3

Vian, M. A., Fernandez, X., Visinoni, F., & Chemat, F. (2008). Microwave hydrodiffusion and gravity, a new technique for extraction of essential oils. *Journal of Chromatography A*, *1190*(1–2), 14–17. 10.1016/j.chroma.2008.02.086

Vilkhu, K., Mawson, R., Simons, L., & Bates, D. (2008). Applications and opportunities for ultrasound assisted extraction in the food industry - A review. *Innovative Food Science and Emerging Technologies*, *9*(2), 161–169. 10.1016/j.ifset.2007.04.014

Vinatoru, M. (2001). An overview of the ultrasonically assisted extraction of bioactive principles from herbs. *Ultrasonics Sonochemistry*, *8*(3), 303–313. 10.1016/S1350-4177(01)00071-2

Zhang, Q. W., Lin, L. G., & Ye, W. C. (2018). Techniques for extraction and isolation of natural products: A comprehensive review. *Chinese Medicine*, 1–26. 10.1186/s13020-018-0177-x

Zhou, Y. J., Xue, C. M., Zhang, S. S., Yao, G. M., Zhang, L., & Wang, S. J. (2017). Effects of high intensity pulsed electric fields on yield and chemical composition of rose essential oil. *International Journal of Agricultural and Biological Engineering*, *10*(3), 295–301. 10.3965/j.ijabe.20171003.3153

# 7 Chemical Composition of Essential Oils

## 7.1 INTRODUCTION

Human civilization has been known to utilize aromatic plants and herbs since around 5000 BC. Aromatic herbs were identified in primary care and used as therapeutic agents in treating several diseases. It was not confined to any geographical area and, based on the trade data and residual patterns in contemporary society, indicated the reported uses of aromatic plants in nearly every part of the world in the healthcare system. Now wider therapeutic applications are being reported (Buckle, 2011). Aromatic herbs and spices then began to be used in various foods not only for flavouring but also for preservative purposes. These plants are considered an untapped reservoir of valuable substances called phytochemicals, phytogenic, phytobiotics, botanicals or spices. However, they are never established as essential ingredients or have gained a significant commercial identity.

## 7.2 BASICS OF ODOURS AND ASSOCIATED COMPLEXITY

The odours or smells we perceive are primarily the results of a mixture of odorants. There can be one or multiple odours associated with an odorant. Several studies have attempted to link odorant physicochemical properties to specific olfactory perception; however, no universal rule has yet determined how and to what extent molecular properties affect odour perception. The sense of smell is fascinating and undeniably powerful. As humans, we use the sense of smell for various reasons, such as to determine if the food quality is good or degrading (rotting), detect toxic gases or chemicals, and be able to transport us to different locations and times in our memory. Compared to the senses of vision, audition, and touch, it is underappreciated and has received less attention in human research. The olfactory system is a complex combination of processes, including olfactory receptor activation in the nose, electrical signal generation, and our brain. Molecules only in the gaseous state can stimulate our sense of smell; thus, to recognize different smells/odours, the molecules are to be released into the air. Such molecules (odorants) can come directly from the air we breathe or from volatile compounds released in our mouths while chewing. Odorants can be classified as usually low-molecular weight (<300 Da), carbon-containing, water-soluble compounds with high vapour pressure ($\geq$ 0.01 kPa at 20 °C) and lipophilic compounds. However, odorants are not confined to carbon-containing compounds, organic and inorganic molecules, or based on the vapour pressure value. The complex link between odorant chemistry and odour perception is central to olfaction research. A single molecule can have more than one odour profile as it may be linked with the perception, or even molecules with

DOI: 10.1201/9781003315384-7

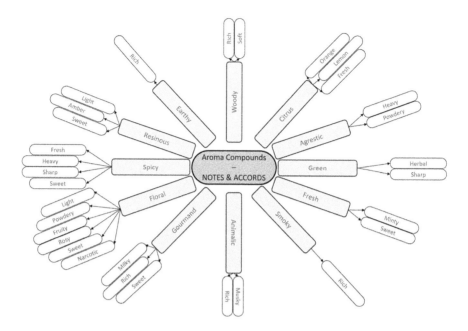

**FIGURE 7.1**    Perceptions of odour profiles based on Notes and Accords.

diverse structural characteristics can have similar odour profiles. Similarly, a slight structural difference can provide a drastic variation or contrasting odours or even change the intensity of the smell. Concluding that a molecule's odour can be predicted using one or more molecular properties or structures is far from simple. Even with several studies related to structure-odour relationships (SORs) and attempts to link odorant physicochemical properties to specific olfactory perception or structural similarities to predict the odour, there is no universal rule that can determine how much molecular properties or chemical structure affects odour quality and perception (Buckle, 2011; Sue Clarke, 2008; Sharma et al., 2022).

Sharma et al. conducted a study to identify important and common features of odorants with seven basic odours (floral, fruity, minty, nutty, pungent, sweet, and woody) to comprehend the complex topic of odours. The results revealed peculiar links between odours, specific molecular properties linked with each set of odorants, and a common spatial distribution of features for odours. A set of 19 structural features was proposed that can be used to assess seven basic odour classifications quickly (Sharma et al., 2022) (Figures 7.1 and 7.2).

## 7.3   COMPOSITION OF ESSENTIAL OILS

Essential oils/natural volatile (NVEOs) are a complex mixture of hundreds of organic compounds mainly consisting of chemicals from families of hydrocarbons (framework of carbon and hydrogen only), typically unsaturated aliphatic. These hydrocarbons tend to have one or more functional groups attached by replacing hydrogen atoms on the hydrocarbons giving the molecule a specific role in

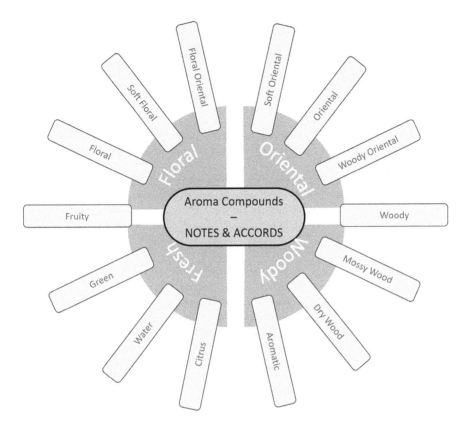

**FIGURE 7.2** Notes and accords of aromatic plants.

intermolecular interaction. Hydrocarbons vary in size and complexity, providing aliphatic compounds (alkanes, alkenes and alkynes), ring or alicyclic structures (mono-, bi, tri-, tetracyclic and so on), or aromatic compounds. The most commonly occurring compounds are terpenes (monoterpene, sesquiterpene, diterpene, tri-terpene and tetra-terpene) derived from active isoprene units, oxygenated hydro-carbons of these terpenes called terpenoids (terpenic alcohols, terpenic aldehydes, terpenic ketones). Other compounds in the mixture are phenols and phenolic ethers, esters, lactones and coumarins (Tisserand & Young, 2014). Most of the constitu-ents, based on the chemical structure, are known to have a unique odour but con-tribute differently to the characteristic smell of the oil as a whole. About 600 compounds are identified under these chemical families (R. P. Adams, 2007; Teixeira et al., 2013).

Some of these are categorized in Table 7.1.

Typically, EOs' composition consists of 2–3 major constituents with the highest concentrations of 20%–70%, 10–25 minor constituents (with less than 1% con-centration) and trace constituents representing 5%–8.5% cumulatively of the whole oil (Sharmeen et al., 2021; Tisserand & Young, 2014). The composition of the

**TABLE 7.1**

**Commonly Seen Compounds in EOs as Per their Chemical Families (Compounds Highlighted in Bold are Prominently Found in Indian EOs) *(S* Clarke, 2008; Hanif et al., 2019; Tisserand & Young, 2014)**

| Family of Compounds | Compounds |
|---|---|
| Terpenes – Monoterpenes | Bornylene, Camphene, Carene, Cymene, Dipentene, Limonene, Menthene, Myrcene, Ocimene, Phellandrene, Sabinene, Terpinene, Terpinolene, Thujene, α-Pinene, α-Terpenene, β-Pinene, γ-Terpenine, 2,6,6-Trimethyl-1-methyl-cyclohex-2-ene, Terpinolene |
| Terpenes – Sesquiterpenes | Aromadendrene, Bergamotene, **Bisabolene, Bourbonene, Bulnesene, Cadinene, Caryophyllene, Cedrene, Copaene,** Cubene, Elemene, **Farnesene,** Germacrene, **Guaiene, Humulene,** Lepidozene, **Longifolene,** Muurolene, Patchoulene, Selinene, Seychellene, Viridiflorene, Ylangene, β- Elemene |
| Terpenes – Diterpenes | Camphorene, Gibberellic acid |
| Alcohols | n-Decanol, 3,7,11,15-tetramethyl-2-hexadecen-ol,A |
| Alcohols – Monoterpeneols | **Borneol,** Citronellol, Fenchyl alcohol, Geraniol, Isopulegol, Lavandulol, Linalool, Myrtenol, Nerol, Terpin-4-ol, Terpineol, Terpinol-1-ol, α-Cadinole, α-Costol, α-Selinenol, Δ-Cadinol, τ-Muurolol, |
| Alcohols – Sesquiterpeneols | Atlantol, Bisabolol, Cadinol, Caryophyllene alcohol, Cedrol, Elemol, Eudesmol, **Farnesol,** Nerolidol, **Patchoulol, Santalol,** Viridiflorol |
| Alcohols – Diterpeneols | Sclareol |
| Aldehydes | Acetaldehyde, **Anisaldehyde,** Benzaldehyde, Caproic aldehyde, **Cinnamaldehyde, Citral,** β-**Citronellal, Cuminaldehyde,** Decanal, Dodecenal, 2-Dodecenal, capric aldehyde, Laural aldehyde, **Geranial,** Myrtenal, **Neral,** Nonanal, Perillaldehyde, Phellandral, Piperonal, Sinensal, Stearaldehyde, Teresantal, Vanillin, Valeranal |
| Ketones | Artemisia ketone, (+) - Camphor, (−) – Carvone, (R)-(−)-Fenchone, α-Ionone, cis-α-Irone, cis-Jasmone, (−)-Menthone, 6-Methylhept-5-en-2-one, 2-Nonanone, Perilla ketone, (+)-Pinocamphone, Pinocarvone, (6S)-(+)-Piperitone, (+)-cis-Pulegone, (+)-b-Thujone, Thymoquinone, ar-Turmerone, 2-Undecanone, (1R,5R)-(+)-Verbenone |
| Acids | Hexadecanoic acid, Terpinyl acetate |
| Oxides | 1,8-cineol, Caryophyllene oxide, α-Bisabolol oxide A, α-Bisabolone oxide A, β-Caryophyllene oxide, 1,8-Cineole, Geranyl oxide, Linalool oxide A, (2R)-Nerol oxide, (−)-cis-Rose oxide, Sclareol oxide |
| Phenols | Carvacrol, Chavibetol, Chavicol, cresol, Eugenol, Hydroxy-chavicol, Iso-eugenol, Thymol |
| Esters | Allylpyrocatecol diacetate, Benzyl benzoate, Bornyl acetate, Bornyl isovalerate, Butyl angelate, Chavibetol acetate, Citronellyl acetate, Citronellyl butyrate, Citronellyl formate, Eugenol acetate, Eugenyl |

**TABLE 7.1 (Continued)**
**Commonly Seen Compounds in EOs as Per their Chemical Families (Compounds Highlighted in Bold are Prominently Found in Indian EOs) (S Clarke, 2008; Hanif et al., 2019; Tisserand & Young, 2014)**

| Family of Compounds | Compounds |
|---|---|
| | acetate, Geranyl acetate, Geranyl tiglate, Hexyl acetate, Lavandulyl acetate, Linalyl acetate, Menthyl acetate, Methyl anthranilate, Methyl benzoate, Methyl butyrate, Methyl salicylate, Neryl acetate, Propyl angelate, Sabinyl acetate, Terpineol acetate, trans-Pinocarveol acetate, Vetiverol acetate |
| Ethers | (E)-Anethole, Dill apiole, β-Asarone, p-Cresyl methyl ether, Elemicin, Estragole (Methyl chavicol), methyl eugenol, (Z)-O-methyl isoeugenol, Phenylethyl methyl ether, Myristicin, Safrole |
| Lactones | Alantolactone, cis-Ambrettolide, Costunolide, Coumarin Dehydrocostus lactone, Massoia lactone, (Z,E)-Nepetalactone (−)-Pentadecanolide |
| Furans | a-Agarofuran, 3-Butyl phthalide, (Z)-Butylidine phthalide, Dihydrobenzofuran, (Z)-Ligustilide, (6R)-(ḃ)-Menthofuran |

volatiles is seen to vary with provinces, subspecies/chemotypes used, distillation period, stage of plant development, plant part used, and other dynamic factors like weather conditions, soil type and fertilizers utilized, time of harvesting, and so on (Chalchat et al., 1997; Özgüven & Tansi, 1998; Piccaglia et al., 1997; Zrira & Benjilali, 1996). Certain examples of major components are as follows; the EO of carvacrol (up to 30%) of thymol (up to 27%) in oraginum species, eucalyptol in eucalyptus (up to 49%), citral in lemongrass (up to 65%), menthol in mint (up to 71%), limonene in lemon oil (up to 68%), bisabolol in german chamomile (up to 38%), myristicin (up to 13.5%) in nutmeg, eugenol in clove (up to 74%), and anethole in star anise (up to 72%). These components determine the biological properties of these aromatic oils, whereas each is included in different groups of distinct biosynthetic origin.

## 7.4 CHEMICAL COMPOUNDS OF ESSENTIAL OILS

As stated above, more than 600 molecules have been identified and reported under about 16 classes of compounds. Typically these are terpenes and phenylpropanoids derivatives with minimal chemical and structural differences. These are mostly volatile compounds like mono-and sesqui-terpene-based components, along with oxygenated molecules like alcohols, aliphatic aldehydes, ketones and esters, and 1%–10% of non-volatile compounds like carotenoids, fatty acids, flavonoids and waxes (Buckle, 2011; Sue Clarke, 2008; Herman, 2017). These compounds can be described below,

### 7.4.1 HYDROCARBONS

Hydrocarbon is a chemical compound found in EOs with building blocks connected by hydrogen and carbon bonding. An example of basic hydrocarbons found in EO oils is isoprene. Mono-, sesqui-, and diterpenes are categorized under the terpenes group.

#### 7.4.1.1 Monoterpenes

These are compounds formed by combining two isoprene units from head to tail. The main properties of monoterpenes are antibacterial, analgesic, stimulant, and expectorant. Monoterpenes are naturally occurring constituents in EO plants, with the majority being unsaturated hydrocarbons (C10).

Alcohols, ketones and carboxylic acids are present as substituents in the oxygenated derivatives of monoterpenes, which are collectively known as monoterpenoids. The branched chains of C10 hydrocarbons, which link two isomers, are widely distributed in herbal plants, with more than 400 naturally occurring monoterpenes identified.

#### 7.4.1.2 Sesquiterpenes

These are formed by joining three isoprene groups together. The structures of sesquiterpenes may be linear, monocyclic or bi- and tricyclic. Linear structures of sesquiterpenes, referred to as farnesenes are branched hydrocarbons with four double bonds. The farnesenes usually found in EOs are E, E-α- farnesene and E-β-farnesene, while E, Z-α-farnesene and Z, Z-α- farnesene are not reported to occur in nature.

Oxygenated products of farnesenes are farnesols as primary alcohol, E- & Z- nerodiol as tertiary alcohol, and α-sinensal & β-sinensal as the aldehydic components. These are typically seen as minor constituents of EOs.

Monocyclic forms of these sesquiterpenes are bisabolene groups (Z-α-bisabolene, β-bisabolene, E-γ-bisabolene, and Z-γ-bisabolene as isomers) with C6 ring structures, α-Zingiberene (ginger EOs), and β-Curcumene & ar-curcumene (Curcuma longa EO).

Bicyclic sesquiterpenes represent a structure like the pinenes, with a cyclobutane ring like bergamotenes (trans-α-bergamotene) seen as a major constituent in Pimpinella affinis and minor constituent in lemon and lodgepole pine (Pinuscontorta) EO. Caryophyllene (E-caryophyllene) derived from α-humulene presents a C9 ring fused to a cyclobutane ring seen as a major component in cannabis, oregano, and rosemary EO.

#### 7.4.1.3 Diterpenes

These are formed by joining four units of isoprene. They are considered too heavy components that do not evaporate easily during the extraction process using steam distillation; hence, it is impossible to be found in isolated aromatic oils. Diterpenes are found in all plant families with C20 chemical structures. Almost 2,500 known diterpenes were discovered, which belong to 20 major structural types. Diterpenic derivatives can be found in plant hormones and phytol, which occur as a side chain on chlorophyll.

## 7.4.2 ALCOHOLS

Alcohols in EOs are termed *mono-terpineol*. As stated earlier, these are mono-terpenes containing hydroxyl groups inside their hydrocarbon structure. The attachment of terpenes with oxygen and hydrogen atoms can result in the formation of alcohols. These may occur as a single compound or combined with a terpene or ester. They are known to contribute some excellent properties like antiseptic, antiviral, antibacterial, and germicidal.

## 7.4.3 ESTERS

The formation of esters is due to the interaction between alcohols and acids. The ester component inside EOs offers soothing and balancing effects. Similar to alcoholic compounds and owing to the alcoholic groups inside esters, they can provide anti-inflammatory activities. In medical sciences, esters are considered to have antifungal and sedative/soothing properties, with balancing action on the nervous system. The common esters found in EOs are linalyl acetate, benzyl acetate, geranyl acetate, citronellyl formate and geranyl formate, which are present in bergamot and lavender, and geranium EOs (Aziz et al., 2018).

## 7.4.4 ACIDS

These are the molecules typically derived from the terpenes or the phenolics containing the carboxyl groups. These are the rare compounds seen in EOs, but there are some cases, like phenylacetic acid seen in neroli EO. Other examples of acids from EOs are anisic acid, vetiveric acid, citronellic acid, cinnamic acid and valerenic acid. Acids are important as they react with the alcohols giving esters.

## 7.4.5 PHENOLS AND PHENOL ETHERS

In EOs, phenols occur as substituted phenols. Some of the functional groups replace one or more of the five hydrogen atoms of the ring. Typical examples of phenolic compounds in EOs are carvacrol, cresol, thymol and eugenol. These are typically seen in spicy herbs like thyme, sage, clove, piper betel, oregano, cinnamon leaf, pimento, ylang-ylang, and so on, contributing spicy, herbaceous, and pungent odour profile. Phenolic compounds contribute antiseptic, anti-infectious and bacterial properties.

Phenolic ethers include safrole, anethole and estragole.

## 7.4.6 ALDEHYDES AND KETONES

Aldehydes have powerful aromas and seek widespread applications in perfumery industries. Typical examples of aldehydes are citronellal, which contributes a strong citrus smell; cinnamic aldehyde, a warm spicy balsamic smell; and geranial, a light, sharp, fresh lemon odour. These highly reactive molecules may oxidize to organic acids if not stored correctly. These contribute properties intermediate between alcohols and ketones, such as anti-infectious, tonic, vasodilator, hypotensive and antipyretic.

Ketones are a majority of EOs and are very common molecules. Typical examples are carvone, menthone, jasmone, camphor, fenchone, inones, thujone and so on. They contribute to the therapeutic properties of digestive, encourage wound healing, and are calming and soothing.

## 7.5  CONCLUSION

Plants are complex structures to study and deconstruct for further processing. Whether dehydration or extraction, the complexity significantly impacts the collection of the desired secondary metabolites. Further to that, plant-based natural volatiles and EOs are complex mixtures of several compounds belonging to the families of terpenes, alcohols, aldehydes, ketones, esters, ethers, and acids, phenolics, alkaloids, lactones and coumarins. A detailed understanding of the composition and these compounds' significance must be understood before proceeding with the processing. Since selective compound in the oil plays a critical role in the respective application, extraction of the same in primitive form is crucial. Hence, whether it be harvesting, dehydration, extraction or direct applications, the associated chemical compounds and roles, perceptivity, volatility, and stability must be considered.

## REFERENCES

Adams, R. P. (2007). *Identification of essential oil components by gas chromatography/mass spectrometry*, 4th Edition , Allured Publ., Carol Stream, IL.
Aziz, Z. A. A., Ahmad, A., Setapar, S. H. M., Karakucuk, A., Azim, M. M., Lokhat, D., Rafatullah, M., Ganash, M., Kamal, M. A., & Ashraf, G. M. (2018). Essential oils: extraction techniques, pharmaceutical and therapeutic potential - A Review. *Current Drug Metabolism, 19*(13), 1100–1110. 10.2174/1389200219666180723144850
Buckle, J. (2011). *Clinical Aromatherapy - Essential oils in Healthcare* (Third edit). Elsevier.
Chalchat, J. C., Garry, R. P., & Michet, A. (1997). Variation of the chemical composition of essential oil of mentha piperita L. during the growing time. *Journal of Essential Oil Research, 9*(4), 463–465. 10.1080/10412905.1997.9700750
Clarke, S. (2008). Families of Compounds That Occur in Essential Oils. In S. Clarke (Ed.), *Essential Chemistry for Aromatherapy* (2nd ed., pp. 41–77). Churchill Livingstone. 10.1016/b978-0-443-10403-9.00003-0
Clarke, S. (2008). *Essential Chemistry of Essential Oils* (4th ed.). Churchill Livingstone, Elseivier.
Hanif, M. A., Nisar, S., Khan, G. S., Mushtaq, Z., & Zubair, M. (2019). Essential Oil Research - Trends in Biosynthesis, Analytics, Industrial Applications and Biotechnological Production. In S. Malik (Ed.), *Essential Oil Research* (1st ed.). Springer. 10.1007/978-3-030-16546-8
Herman, S. (2017). Chapter 18 Fragrance. In *Cosmetic Science and Technology: Theoretical Principles and Application* (pp. 267–283). Elsevier Inc.
Özgüven, M., & Tansi, S. (1998). Drug yield and essential oil of Thymus vulgaris L. as in influenced by ecological and ontogenetical variation. *Turkish Journal of Agriculture and Forestry, 22*(6), 537–542. 10.3906/tar-96123

Piccaglia, R., Marotti, M., & Dellacecca, V. (1997). Effect of planting density and harvest date on yield and chemical composition of sage oil. *Journal of Essential Oil Research*, *9*(2), 187–191. 10.1080/10412905.1997.9699457

Sharma, A., Kumar, R., & Varadwaj, P. K. (2022). Decoding seven basic odors by investigating pharmacophores and molecular features of odorants. *Current Bioinformatics*, *17*(8), 759–774. 10.2174/1574893617666220519111254

Sharmeen, J. B., Mahomoodally, F. M., Zengin, G., & Maggi, F. (2021). Essential oils as natural sources of fragrance compounds for cosmetics and cosmeceuticals. *Molecules*, *26*(3). 10.3390/molecules26030666

Teixeira, M. A., Rodríguez, O., Gomes, P., Mata, V., & Rodrigues, A. E. (2013). Perfume Engineering: Design, Performance and Classification. In *Perfume Engineering: Design, Performance and Classification*. 10.1016/C2012-0-03624-1

Tisserand, R., & Young, R. (2014). Essential Oil Composition. In *Essential Oil Safety* (2nd ed.). © 2014 Robert Tisserand and Rodney Young. 10.1016/b978-0-443-06241-4. 00002-3

Zrira, S. S., & Benjilali, B. B. (1996). Seasonal changes in the volatile oil and cineole contents of five eucalyptus species growing in morocco. *Journal of Essential Oil Research*, *8*(1), 19–24. 10.1080/10412905.1996.9700548

# 8 Future Research Opportunities

## 8.1 INTRODUCTION

Medicinal and aromatic plants are a consistent part of the natural biodiversity endowment across the globe. The secondary metabolites (essential oils, pharmaceuticals, colourants, dyes, cosmetics and biocides) from these plants are region specific. They are important for the well-being of the local people as they significantly improve the body systems' functioning. These plants are also crucial for uplifting local economies, cultural integrities, and building a sustainable ecology. Researchers and conservationists have increasingly acknowledged this fact over the last decades. Progress in pharmacognosy and highlighted benefits of these plants have led to unsustainable exploitation. Only recently, efforts have been put in and prioritized for understanding the distribution, genetic biodiversity, conservation and ecology (Phondani et al., 2011). The high demand for natural isolates has also promoted environmentally friendly agricultural cultivation and participation of local communities. The increase in the involvement of systematic cultivation and high harvest volumes are being seen as, or soon will need to be looked upon as, opportunities for setting and meeting research goals in the domain of processing these medicinal and aromatic plants.

Considering the increase in industrial demand towards sustainable and renewable resources for fine chemicals and the increasing interest of many farmers towards commercial cultivation and primary processing for value addition, the role of R&D is crucial for bridging the gap. In such a scenario, for industrial partners, cultivators and growers, it is important to have a mutual agreement and understand the buyers, product specifications by the buyers, trade regulations, cultivation practices (indoor or outdoor), processing protocols (on-field or centralized processing), optimum processing parameters, storage and transport requirements (Pandey et al., 2019; Pise & Thorat, 2022a).

## 8.2 R&D NEEDS – UNDERSTANDING THE HISTOLOGY AND DRYING

Nature never compartmentalizes in different sectors such as physics, chemistry, biology, engineering, pharmacology and so on. Every topic being studied and explored can be seen from every discipline's point of view to give the research the required completeness. Having said so, it is crucial to explore the plant organization from all domains, even for the dehydration of the aromatic plant material.

Histological exploration and histochemical investigation of various aromatic plants for the plant part significant for contributing the desired phytochemicals are

DOI: 10.1201/9781003315384-8

well reported. However, being a cross-disciplinary subject, exploration of plant organization, especially for facilitating drying and extraction unit operation, is required from an engineering perspective. Micro-level investigation of plant material, especially food and vegetables, for increasing the shelf life is being explored and reported to optimize the product's quality. A detailed understanding of mass and heat transfer from the plant material is also being explored to understand the phenomenon at multiscale (Purlis et al., 2021). Indeed, these explorations are supportive, and analogical studies can help understand the intricacies of drying aromatic plants. However, compared to fruits and vegetables with significant homogenous tissue structures, aromatic plant parts are more heterogeneous. Secretory cells containing the secondary metabolites or the osmophore, trichomes, ducts, and cavities are delicate and surrounded by guard cells, increasing the complexity of structural shrinkage during dehydration. Such studies are still required at fundamental and applied levels to understand the natural matrix better. In addition, considering the uniqueness of different plant organs and different aromatic plants, detailed classification based on similarities and characterization of the properties will help develop generic models and protocols for the dehydration and deconstruction of plant material to improve the quality of the required products.

## 8.3   R&D NEEDS – UNDERSTANDING OF DEHYDRATION PARAMETERS AND TECHNOLOGY FOR MULTIMODE DRYER

The benefits of different dryers in the chemical industry are well established over decades of research. However, the benefits of various dryers for drying agricultural products is yet being explored (Delele et al., 2015; Oztekin & Martinov, 2007). Still, the most widely used practice for dehydration of agricultural produces is hot air high-flow dryers. More studies on optimizing the design and operation are needed to exploit different heating modes, types of contacting, and operating conditions optimization. Investigations on agricultural products like cereals and legumes are explored to a larger extent and reported, followed by fruits and vegetables, spices and herbs. However, considering the upcoming industrial demands and scope of exploration, there is a significant research requirement from the micro-level to bulk-level dehydration and its impact on the desired compound retention.

Extensive experimental and mathematical modelling techniques can be undertaken considering the enormous biological variability of aromatic plants and their thermolabile nature. Validated mathematical models based on physical, chemical and biological principles are becoming viable alternatives. The numerical methods of computational fluid dynamics (CFD) and discrete element method (DEM) can be explored and implemented to investigate compartment processes such as particle flow, airflow, heat and mass transfer (Mutuli et al., 2020). The studies on the thermo-sensitivity of aromatic plants; the structural organization of targeted tissues; and the relation between moisture content, water activity and moisture equilibrium curve for the plant material will provide insight into the stabilization of the material under different conditions. This will help define a multistage drying protocol and systematically use different drying modes for efficient operations (Khaing Hnin et al., 2019).

## 8.4   R&D NEEDS – UNDERSTANDING OF KEY COMPONENTS AND SPECIFIC EXTRACTION

Medicinal and aromatic plants are important for the secondary metabolites secreted by them. The cultivation, harvesting, pre-processing market demand and global trade are only given significant importance because of these secondary metabolites. As stated in the previous section, commercial cultivation of these plants at an industrial scale and low content of desired natural volatiles to be extracted gives rise to efficient and selective extraction technology for processing the herbage. To reduce intermittent and high-volume processing requirements of extraction plants, it is important to consider dehydration of the plant material so that the metabolites are stored in the natural matrix with the least alteration in the composition (Pise & Thorat, 2022a). However, the separation of the desired natural volatiles and es-sential oils in processible form is still an essential operation and needs significant exploration to be carried out.

Though the extraction process is well-reported and the technologies for extrac-tion, including hydro- or steam-distillation, solvent extraction, and supercritical fluid extraction, are well-established, the extraction process is rarely implemented efficiently. Massive quantities of essential oils are still extracted by steam distil-lation on fields, with operators having a lack of understanding of the process. The equipment implemented for extraction is crudely designed without giving due consideration to critical parameters like an actual steam requirement, steam biomass contacting, time for extraction and texture and structure of the plant material being exposed to extraction. A detailed and standard set of studies needs to be carried out and reported in simple operating protocols for the farm level extraction of desired components on various aromatic plant species. The set of information and protocol can be defined in the form of an SOP considering sorting and utilizing specific parts for extraction, packing density, stacking protocol, consideration of inserts for uni-form steam distribution, the temperature profile of the extraction unit (at the steam inlet, biomass bed, steam outlet, condenser, and hydrosol collection temperature), steam pressure at the generator, temperature ramp up time, extraction kinetics & time for extraction, steam flowrate and oil carrying capacity of the steam. In addition to designing and distribution of modular, low-cost, efficient farm distil-lation units, studies need to be carried out on the storage and shelf life of the essential oils collected. As the oils are of importance because of the composition of the oils and only a few components, a detailed understanding of the actual com-position, losses during extraction, storage & transit, and change in composition owing to oxidation needs to be studied and reported for all the commercially important oils.

Though solvent extraction is widely implemented at the industrial scale, supercritical fluid extraction is a cleaner, greener and more efficient process gaining significant popularity. Detailed studies and a standard set of results reporting ex-traction kinetics and impacts of different variable extraction parameters like pres-sure, temperature, solvent flow rate, solvent polarity and time for extraction can provide insight into the enhancement of the process (Pise & Thorat, 2022b). Presently, rectification of the extracted components is a separate process. However, strategizing and conducting experiments using experimental design and analysis tools will improve the product realization process. Linking the process variables and the polarity of extraction solvents with the composition of the extract will help reduce the rectification process and lead to a single-stage selective extraction protocol.

The use of a co-solvent in supercritical fluid extraction has resulted in better extraction and contributed to the understanding of the extraction fundamentals (Ribeiro Grijó et al., 2019). The promising future developments in supercritical fluid extractions will help establish multistage extraction processes for obtaining the complete extract with the desired separation. Furthermore, the plant volatiles being collected presently through known extraction processes are from the secretory glands within the tissues. These may not be the actual composition secreted by the plants, as some compounds may be lost due to volatility during harvesting and processing. Headspace analysis gives an insight into the composition of the aroma being sensed by the aromatic plants (Erbaş & Baydar, 2016). However, a major challenge lies in developing novel model systems for performing the experimentation and preserving the components in natural conditions.

## REFERENCES

Delele, M. A., Weigler, F., & Mellmann, J. (2015). Advances in the application of a rotary dryer for drying of agricultural products: A review. *Drying Technology*, *33*(5), 541–558. 10.1080/07373937.2014.958498

Erbaş, S., & Baydar, H. (2016). Variation in scent compounds of oil-bearing rose (Rosa damascena Mill.) produced by headspace solid phase microextraction, hydrodistillation and solvent extraction. *Records of Natural Products*, *10*(5), 555–565.

Khaing Hnin, K., Zhang, M., Mujumdar, A. S., & Zhu, Y. (2019). Emerging food drying technologies with energy-saving characteristics: A review. *Drying Technology*, *37*(12), 1465–1480. 10.1080/07373937.2018.1510417

Mutuli, G. P., Gitau, A. N., & Mbuge, D. O. (2020). Convective drying modeling approaches: A review for herbs, vegetables, and fruits. *Journal of Biosystems Engineering*, *45*(4), 197–212. 10.1007/s42853-020-00056-9

Oztekin, S., & Martinov, M. (2007). *Medicinal and Aromatic Crops - Harvesting, Drying, and Processing* (S. Oztekin & M. Martinov (eds.)). Haworth Forr and Agricultural product Press.

Pandey, A. K., Kumar, P., Saxena, M. J., & Maurya, P. (2019). Distribution of aromatic plants in the world and their properties. In *Feed Additives: Aromatic Plants and Herbs in Animal Nutrition and Health* (pp. 89–114). Elsevier Inc. 10.1016/B978-0-12-814 700-9.00006-6

Phondani, P. C., Negi, V. S., Bhatt, I. D., Maikhuri, R. K., & Kothyari, B. P. (2011). Promotion of medicinal and aromatic plants cultivation for improv- ing livelihood security: A case study from West Himalaya, India. *Int. J. Med. Arom. Plants*, *1*(3), 1–8. http://www.openaccessscience.com

Pise, V. H., & Thorat, B. N. (2022a). Role of Drying Process on Quality of Essential Oils and in F & F Industry. *IDS 2022, International Drying Symposium Proceeding*, *1*. 10.55 900/fztcaezx

Pise, V. H., & Thorat, B. N. (2022b). Supercritical fluid extraction of dried Surangi flowers ( Mammea suriga). *Industrial Crops & Products*, *186*(April), 115268. 10.1016/ j.indcrop.2022.115268

Purlis, E., Cevoli, C., & Fabbri, A. (2021). Modelling volume change and deformation in food products/processes: An overview. *Foods*, *10*(4), 778. 10.3390/foods10040778

Ribeiro Grijó, D., Vieitez Osorio, I. A., & Cardozo-Filho, L. (2019). Supercritical extraction strategies using CO2 and ethanol to obtain cannabinoid compounds from cannabis hybrid flowers. *Journal of CO2 Utilization*, *30*(January), 241–248. 10.1016/j.jcou.2018.12.014

# Annexure I

## List of Aromatic Plants (Lodha & Telang, 2016; Thomas et al., 2016)

| Sr. No. | Item | Scientific Name | Family | Part Utilised | % MC in Biomass | Hydro-/ Steam | Solvent | Super Critical | Known Compounds | Major Compounds/Key Compounds | Applications | Reference |
|---|---|---|---|---|---|---|---|---|---|---|---|---|
| 1 | Musk-dana / Lata-kasturi / Kasturbhed | *Abelmoschus moscatus* | Malvaceae | Seeds | 7%–10% | 0.15%–0.2% | 3.20% | 7.9% | 27 | Ambrettolide, Farnesyl acetate, Decyl acetate, Farnesol | Perfumery and medicinal application, flavouring food | (Rout et al., 2004) |
| 2 | Calamus | *Acorus calamus* | Acoraceae | Herb & rhizome | 68%–70% | 1.80% | NA | 3.50% | 32 | α-asarone, (E)-methyl-isoeugenol, and methyl-eugenol. | Medicinal applications, insecticidal and repellent | (Liu et al., 2013; Marongiu et al., 2005) |
| 3 | Galangal | *Alpinia calcarata* | Zingi-beraceae | Rhizome | 75%–78% | 0.25%–0.3% | NA | NA | 49 | Cineole | Medicinal applications and aphrodisiac | (Rahman & Islam, 2015) |
| 4 | Dill | *Anethum sowa* | Apiaceae | Herb, seed | 20%–24.5% | 3.20% | 5.24% | 4.11% | 24 | Dillapiol, m-diamino-benzene, β-butyrolactone | Seasoning pickles, aroma chemicals, pharmaceuticals | |
| 5 | Celery / Bodiaja-moda | *Apium graveolens* | Apiaceae | Herb | 86.00% | 0.25%–0.6% | 0.90% | NA | 37 | Limonene, neo-phytadiene, sedanen-olide, neocnid-ilide | Flavouring foods, pharmaceuticals | |
| 6 | Davana | *Artemisia pallens* | Asteraceae | Flowering tops | 80.00% | 0.20% | NA | NA | 26 | Davanone, fenchyl alcohol, davanofuran | Natural flavours, beverages, high-grade perfumery | |
| 7 | Linaloe | *Bursera delpechiana* | Burseraceae | Wood, berries, leaves | 72%–80% | 2.16% | NA | NA | 15 | Linalool, linalyl acetate, methyl heptanol | Aroma chemicals, perfumery | |
| 8 | Ylang ylang/ Cananga | *Cananga oderata* | Annonaceae | Flower | 62%–65% | 1.1%–2.5% | 3.80% | NA | 43 | Cineole, Linalool, methyl-anisole, methyl benzoate, benzyl acetate, and geranyl acetate | Perfumery, soaps, detergents, shampoo | |

| 9 | Paprika | *Capsicum annum* | Solanaceae | Fruit | 75.8%–81.3% | 0.48% | 9.40% | 10.60% | 140 | Capsanthin, hexyl iso-pentanoate, hexyl 2- methyl-butanoate, hexyl iso-hexanoate | Seasoning, blending and natural flavours |
| 10 | Caraway | *Carum carvi* | Apiaceae | Seeds | 7%–10% | 1.68% | 4.20% | 3.94% | 30 | Carvone, limonene | Flavouring |
| 11 | Cedarwood | *Cedrus deodara* | Pinaceae | Wood, sawdust, root | 19%–34.8% | 0.8%–1% | 6.80% | NA | 40 | Cedrol, cedryl acetate, himachalene | In perfumery for fixative effects and unique odour, soaps |
| 12 | Roman chamomile | *Chamaemelum nobile* | Asteraceae | Flower | 83%–85% | 0.4%–1.9% | 3.50% | 3.81% | 14 | Isobutyl angelate, 2-methyl butyl angelate, propyl tiglate, isoamyl angelate and 3-methylbutyl isobutyrate | Flavouring foods, cosmetics and aromatherapy |
| 13 | German chamomile/ Rasna | *Chamomilla recutita / Matricaria chamomilla* | Asteraceae | Flower & herb | 83%–85% | 0.82% | 3.50% | 3.81% | 55 | α-bisabolol oxide, Chamazulene, azulene, farnesene | Flavouring foods, cosmetics, aromatherapy nd in medicinal |
| 14 | Camphor | *Cinnamomum agasthyama-layanum* | Lauraceae | Leaves | 35%–75% | 3.60% | NA | NA | 42 | Camphor, Limonene, α-Pinene, α-Terpineol, Methyl eugenol | Relieve muscle spasm, mild mucolitic property and can reduce bronchospasm |
| 15 | Camphor | *Cinnamomum camphora* | Lauraceae | Wood, leaves | 35%–75% | 3.2%–3.35% | 6.70% | 4.31% | 33 | Safrole, piperitone, sabinene, eugenol | Pharmaceuticals, incense, balms |
| 16 | Tejpat | *Cinnamomum tamala* | Lauraceae | Leaves | 35%–55% | 0.35%–0.72% | NA | 4.44% | 20 | Methyl eugenol, eugenol, trans-cinnamyl acetate and beta-caryophyllene | Flavouring |
| 17 | Dieng-lorthia/ Dalchini | *Cinnamomum pauciflorum* | Lauraceae | Leaves | 35%–55% | 1.36% | NA | 4.55% | 25 | Cinnamal-dehyde | Flavouring agent and in perfumery |

*(Continued)*

| Sr. No. | Item | Scientific Name | Family | Part Utilised | % MC in Biomass | Hydro-/Steam | Solvent | Super Critical | Known Compounds | Major Compounds/Key Compounds | Applications | Reference |
|---|---|---|---|---|---|---|---|---|---|---|---|---|
| 18 | Srilanka cinnamon | *Cinnamomum zeylanicum* | Lauraceae | Leaves, bark & fruit | 19%–35% | 0.70% | 0.72% | 3.80% | 25 | (E)-cinnamal-dehyde, benzaldehyde, (E)-cinnamyl acetate, limonene and eugenol | Flavouring agent and in perfumery | (Unlu et al., 2010) |
| 19 | Lime | *Citrus aurantifolia* | Rutaceae | Fruit peel | 74%–78% | 1.00% | 2.18% | 1.24% | 55 | Limonene, terpineol, cineole, pinene, cymene, bisabolene, citral | Flavouring agent and in perfumery | (Spadaro et al., 2012) |
| 20 | Cassia / Tejpat | *Cinnamon cassia* | Lauraceae | Bark, leaf | 35%–55% | 2%–2.5% | 6.58% | 1.74% | 31 | Eugenol, iso-eugenol, methyl eugenol | Flavouring | |
| 21 | Cinnamon | *Cinnamon verum* | Lauraceae | Bark, leaf | 35%–55% | 1.4%–1.5% | 6.83% | 7.80% | 25 | Eugenol, iso-eugenol, methyl eugenol | Seasoning, blending, natural flavours, pharmaceuticals | |
| 22 | Bergamot | *Citrus bergamia* | Rutaceae | Fruit peel | 74%–78% | 0.24%–1.2% | NA | 22.3% DB | 55 | Linalyl acetate, linalol, limonene | Citrus soft drinks flavour, "East Gray" tea flavour | |
| 23 | Lemon / Citrus / Limbu | *Citrus limon* | Rutaceae | Fruit peel | 74%–78% | 0.4%–1.22% | – | 1.36% | 21 | β-Pinene, limonene, linalool, terpinol and linalyl acetate | Manufacture of soaps, talcs, toothpastes, shaving creams, perfumes, skin care products | |
| 24 | Sweet orange / Santra | *Citrus sinensis* | Rutaceae | Fruit peel | 74%–78% | 0.08%–1.2% | 5.40% | NA | 24 | Citral, limonene | Lemon flavour, seasoning, cosmetic products | |
| 25 | Coriander / Dhane / Kothimbir | *Coriandrum sativum* | Apiaceae | Herb & seed | 82%–84% | 0.1%–0.6% | 14.45% | 8.88% | 44 & 53 | Linalool, α-pinene, phellandrene, camphor | Flavouring, seasoning and curry blends | |

| | Common Name | Botanical Name | Family | Part | | | | | | Major Constituents | Flavour & fragrance uses |
|---|---|---|---|---|---|---|---|---|---|---|---|
| 26 | Saffron / Keshar | *Crocus sativus* | Iridaceae | Flower | 12%–18% | 0.413%–1.00% | 1.31% | 1.03% | 44 | Safranal | Flavour & fragrance industry and traditional medicine |
| 27 | Cumin / Jeera | *Cuminum cyminum* | Apiaceae | Seeds | 7%–22% | 3.16% | 15.60% | 5.24% | 45 | Cuminyl alcohol, cuminal-dehyde | Seasoning curries, natural flavours and perfumery |
| 28 | Turmeric | *Curcuma longa* | Zingiberaceae | Rhizomes | 78%–90% | 0.46%–1.1% | 5.49% | 6.98% | 26 | Turmerone, terpenolene, zingiberne, curcumine | Seasoning, blending, natural flavours, pharmaceuticals |
| 29 | Lemongrass/ Gavati Chaha | *Cymbopogon flexuosus/ citratus* | Poaceae | Grass / leaves | 69%–72% | 0.9%–1% | NA | NA | 27 | Citral, linalool, geraniol | Lemon flavour, seasoning, cosmetics, medicinal industry and perfumes |
| 30 | Palmarosa / Rosha grass | *Cymbopogon martinii* | Poaceae | Flowering tops & leaves | 65%–70% | 0.75%–0.9% | – | – | 32 | Geraniol, geranyl acetate, citronellol, linalool | Flavouring tobacco, soap, high grade perfumery |
| 31 | Citronella Ceylon | *Cymbopogon nardus* | Poaceae | Grass / leaves | 69%–72% | 0.80% | NA | NA | 35 | Citronellal, geraniol, citronellol, geranyl acetate | Perfumery chemicals, soap, cosmetics, flavouring |
| 32 | Citronella Java | *Cymbopogon winterianus* | Poaceae | Grass / leaves | 69%–72% | 0.65%–0.8% | | 2.2% DB | 23 | Citronellal, geraniol, citronellol, geranyl acetate | Perfumery chemicals, soap, cosmetics, flavouring |
| 33 | Elaichi cardamom/ Veldode | *Elettaria cardamomum* | Zingiberaceae | Seeds | 80%–85% | 5% | 7.60% | 5.50% | 67 | Cineole, terpineol, limonene, cymene | Flavouring agent |
| 34 | Eucalyptus | *Eucalyptus citriodora* | Myrtaceae | Leaves & twigs | 35%–55% | 0.70% | | 0.91% | 55 | Citronellal, citronellol, cineole, iso-pulegol | Disinfectants, germicides, soap, cosmetics |

*(Continued)*

| Sr. No. | Item | Scientific Name | Family | Part Utilised | % MC in Biomass | Hydro-/Steam | Solvent | Super Critical | Known Compounds | Major Compounds/Key Compounds | Applications | Reference |
|---|---|---|---|---|---|---|---|---|---|---|---|---|
| 35 | Eucalyptus / Nilgiri | *Eucalyptus globulus* | Myrtaceae | Leaves & twigs | 35%–55% | 1–2% | 2.20% | 3.60% | 39 | Cineole, caryophyllene, camphene, sabinene, myrcene, | Blending medicinal | (Sharafan et al., 2022) |
| 36 | Clove | *Eugenia caryophyllus* | Myrtaceae | Bud, leaf, stem | 28%–30% | 2.2%–2.5% | 4.80% | NA | 18 | Eugenol, caryophyllene, humulene, Hinesol | Seasoning, blending, natural flavours, pharmaceuticals | |
| 37 | Hing | *Ferula assafoetida* | Apiaceae | Rhizomes | 78%–82% | 1.13% | NA | 4.31% | 30 | Germacrene, (E)-1-propenyl sec-butyl disulfide | Flavouring agent | |
| 38 | Fennel / Badishep | *Foeniculum vulgare* | Umbelli-ferae | Fruit | 7%–22% | 1.74% | NA | 2.20% | 28 | Anethole, fenchone, Trans-anethole and estragole | Seasoning, blending, natural flavours, pharmaceuticals | |
| 39 | Star anise | *Illicium verum* | Schisan-draceae | Fruit & seed | 65%–68% | 4.20% | 9.30% | 9.20% | 60 | Trans-anethol, estragole, chalcone | Natural flavours, pharmaceuticals and therapeutic | |
| 40 | Jasmine | *Jasminum grandiflorum* | Oleaceae | Flowers | 70%–95% | 0.05% | 0.35% | 0.26% | 100 | Linalool, benzyl alcohol, indole, cis-jasmone, geraniol, methyl anthramilate | | |
| 41 | Jasmine / Mograa | *Jasminum sambac* | Oleaceae | Flower | 70%–95% | NA | 0.334% Concrete 0.021% Absolute | 0.13% | 53 | Benzyl acetate, linalool, linayl acetate, jasmone | Perfumery, cosmetic ndustry and soap making. Skin toner and conditioner | (Wei et al., 2015) |
| 42 | Jasmine | *Jasmium officinale* | Oleaceae | Flower | 70%–95% | 0.12%–0.13% | 0.34% | 0.13% | 40 | Benzyl acetate, linalool, linayl acetate, jasmone | Natural flavours | |
| 43 | Juniper berry | *Juniperus communis* | Cupressa-ceae | Berries, barks and roots | 55%–80% | 0.26% | – | 0.70% | 70 | α-Pinene, myrcene, sabinene, limonene and β-pinene | Aromatherapy and natural antiseptic | |

| No. | Common name | Botanical name | Family | Part used | | | | | | Chemical constituents | Uses | Reference |
|---|---|---|---|---|---|---|---|---|---|---|---|---|
| 44 | Nameka- cholam / Kachri | *Kaempferia galanga* | Zingi- beraceae | Rhizomes | 78%–82% | 0.6%–0.9% | NA | NA | 42 | ethyl-trans-p-methoxy cinnamate, pentadecane, 1,8-cineole, -carene and borneol | Aromatic, flavour and perfumery as well as aromatherapy | |
| 45 | Kacholam | *Kaempferia rotunda* | Zingi- beraceae | Rhizomes | 78%–82% | 2.4%–3% | NA | NA | 75 | Ethyl-trans-p-methoxy cinnamate, pentadecane, cineole, carene, borneol | Medicinal, preservation | |
| 46 | Bay laurel leaf | *Laurus nobilis* | Lauraceae | Leaves | 35%–75% | 0.90% | – | 0.82% | 29 | 1,8-cineole, 1-(S)-α-pinene, and R-(+)- limonene | Soap making, perfumery, food and cosmetic industry | |
| 47 | Lavender / Dharu | *Lavandula angustifolia* | Lamiaceae | Flower | 58%–72% | 0.99% | 2.4%–5% | 3.70% | 34 | Linalool, linalyl acetate | Herbal medicine, perfumery, soap- making, aromatherepy | |
| 48 | Lavandin | *Lavandula hybrida* | Lamiaceae | Flower | 58%–72% | 1.52% | 2.1%–4% | 1.43% | 42 | Linalool, linalyl acetate | Perfumery, soap, antiseptic, insecticides | |
| 49 | Champaca / Chafa | *Magnolia champaca* | Magno- liaceae | Flower | 70%–95% | 0.03% | EO – 0.03% Concrete -1.25%- 1.75% | 1.04% | 87 | Cinole, iso-eugenol | Perfumery, cosmetics and massage oils | |
| 50 | Cajeput | *Melaleuca Leucadendron* | Myrtaceae | Leaves | 35%–75% | 1.28% | – | 4.11% | 26 | Cineol, eudesmol, viridiflorol, guaiol, terpineol, limonene | Medicinal & expectorant | (Kumar et al., 2005) |
| 51 | Japanese mint / Pudina | *Mentha arvensis* | Lamiaceae | Leaves & twigs | 88%–92% | 0.45%–0.55% | NA | NA | 36 | Menthol, menthone, terpenes | Flavouring tooth pastes, andies, ointments, obacco, cough syrups | |
| 52 | Peppermint | *Mentha piperita* | Lamiaceae | Leaves & twigs | 88%–92% | 0.32%–0.65% | NA | NA | 42 | Menthol | Flavouring, medicinal | |

*(Continued)*

| Sr. No. | Item | Scientific Name | Family | Part Utilised | % MC in Biomass | Hydro-/Steam | Solvent | Super Critical | Known Compounds | Major Compounds/Key Compounds | Applications | Reference |
|---|---|---|---|---|---|---|---|---|---|---|---|---|
| 53 | Spearmint | Mentha spicata | Lamiaceae | Leaves & twigs | 88%–92% | 0.32%–0.65% | NA | NA | 45 | Menthol | Chewing gum, oral hygiene | |
| 54 | Curry leaf / Kadipatta | Murraya koenigii | Rutaceae | Leaves & twigs | 80%–88% | 1.2%–2.5% | NA | NA | 90 | α-Pinene, β-pinene, β-phellandrene, (E)-Caryophyllene and α-selinene | Flavour and perfumery, soap and cosmetic industry, massage oils, aroma therapy | |
| 55 | Nutmeg / Jayphal | Myristica fragrans | Myristicaceae | Fruit, seed, aril, leaf | 12%–20% | 1.40% | 9.63% | 14.40% | 51 | Trimyristin, pinene, camphene, myristicin | Flavour, meat seasoning, baking, perfumery and pharmaceutical industries | |
| 56 | American basil | Ocimum americanum | Lamiaceae | Herb | 88%–92% | 0.22%–0.45% | NA | NA | 101 | Methyl chavicol, citral, linalool | Pharmaceuticals, aroma chemicals, ointments, balms | |
| 57 | French basil | Ocimum basilicum | Lamiaceae | Herb | 88%–92% | 0.21%–0.5% | – | 2% DB | 101 | Methyl chavicol, methyl cinnamate, eugenol, linalool | Pharmaceuticals, aroma chemicals, ointments, balms | |
| 58 | Hoary basil | Ocimum canum | Lamiaceae | Herb | 88%–92% | 0.2%–0.3% | NA | NA | 101 | Linalool, camphor | Pharmaceuticals, aroma chemicals, ointments, balms | |
| 59 | Clocimum | Ocimum gratissimum | Lamiaceae | Leaves & twigs | 88%–92% | 0.3%–0.63% | NA | NA | 101 | Eugenol, methyl chavicol, methyl cinnamate, linalool | Pharmaceuticals, aroma chemicals, ointments, balms | |

| No. | Common name | Botanical name | Family | Part | % | % | % | % | No. | Constituents | Uses | Reference |
|---|---|---|---|---|---|---|---|---|---|---|---|---|
| 60 | Camphor basil | *Ocimum kilimanscharicum* | Lamiaceae | Leaves & twigs | 88%–92% | 0.32%–0.74% | NA | NA | 101 | Camphor | Pharmaceuticals, aroma chemicals, ointments, balms | |
| 61 | Ocimum / Tulsi | *Ocimum sanctum* | Lamiaceae | Leaves & twigs | 88%–92% | 0.45%–0.65% | NA | NA | 101 | Eugenol, methyl chavicol, methyl cinnamate, linalool | Pharmaceuticals, aroma chemicals, ointments, balms | |
| 62 | Holy / Sacred basil | *Ocimum tenuiflorm* | Lamiaceae | Leaves & twigs | 88%–92% | 0.1%–0.2% | NA | NA | 101 | Eugenol, linalool | Pharmaceuticals, aroma chemicals, ointments, balms | |
| 63 | Ocimum | *Ocimum viride* | Lamiaceae | Leaves & twigs | 88%–92% | 0.2%–0.74% | NA | NA | 101 | Thymol | Pharmaceuticals, aroma chemicals, ointments, balms | |
| 64 | Marjoram | *Origanum majorana* | Lamiaceae | Leaves & twigs | 88%–92% | 0.80% | 9.10% | 3.80% | 35 | Terpiniol, sabinene hydrate, a-terpinene, sabinene | Antimicrobial agent | (Busatta et al., 2008) |
| 65 | Kewada | *Pandanous odora-tissimus* | Panda-naceae | Flower | 76%–80% | 0.22%–0.052-% | 11.48% | NA | 59 | 2-Phenyl ethyl methyl ether, terpinen-4-ol, geraniol, β-caryophyllene, β-gurjunene, γ-muurolene and leden | Flavouring, perfumes, Paan masala, lotions, tobacco products, hair oil, cosmetics, soaps, incense sticks | |
| 66 | Geranium / Rose geranium. | *Pelargonium graveolens* | Gerania-ceac | Leaves & twigs | 85%–88% | 0.07%–0.19% | NA | 2.53% | 28 | Geraniol, citronellol, linalool, iso-menthone | Perfumery, soap, cosmetic, flavouring and aromatherapy | |
| 67 | Pimenta / Allspice | *Pimenta diocia* | Myrtaceae | Leaves, twigs & berries | 50%–66% | 1.4%–3.2% | 6.80% | 6.40% | 45 | Eugenol, alpha-pinene | Seasoning, blending, natural flavours | (Pérez-Alonso et al., 2011) |

*(Continued)*

| Sr. No. | Item | Scientific Name | Family | Part Utilised | % MC in Biomass | Hydro-/ Steam | Solvent | Super Critical | Known Compounds | Major Compounds/Key Compounds | Applications | Reference |
|---|---|---|---|---|---|---|---|---|---|---|---|---|
| 68 | Bayleaf | *Pimenta racemosa* | Myrtaceae | Leaves | 35%–55% | 2.50% | NA | 4% | 29 | 1, 8-Cineol, Eugenol, alpha-pinene | Food flavouring, fragrance and aromatherapy | (García et al., 2002) |
| 69 | Anise | *Pimpinella anisum* | Apiaceae | Fruit | 25%–80% | 1.5%–6% | 15.40% | 7.70% | 19 | Anethol, Methyl chavicol | Flavouring soft drinks, confectionery, pharmaceuticals | |
| 70 | Black pepper/ Kali mirch/ miri | *Piper nigrum* | Piperaceae | Seeds | 55%–80% | 0.6%–2.1% | 2.88% | 2.16% | 51 | Piperene | Seasoning, blending, natural flavours, pharmaceuticals | |
| 71 | Patchouli | *Pogostemon cablin* | Lamiaceae | Leaves & twigs | 86%–89% | 1.50% | 2.87% | 5.07% | 17 | Patchoulinol, caryophy-llene | Flavouring non-alcoholic beverages, perfumes, soaps, cosmetics | |
| 72 | Tuberose/ Nishi-gandh. | *Polianthus tuberosa* | Amarylli-daceae | Flower | 70%–95% | 0.08%–0.11% | 0.14% | NA | 37 | Geraniol, nerol, farnesol | Perfumery, cosmetics, aphrodisiac, relaxing, sedative | |
| 73 | Rose | *Rosa damascena* | Rosaceae | Flower | 70%–95% | 0.05% | 0.25% | NA | 32 | Citronellol, geraniol, nerol, linalool | Perfumery, cosmetics, flavouring soft drinks, pharmaceuticals | |
| 74 | Rosemary | *Rosmarinus officinalis* | Lamiaceae | Leaves & twigs | 75%–80% | 0.35% | NA | 1.75% | 30 | Pinene, borneol, camphene, camphor, verbenone | Seasoning blends, flavoring in beverages, cosmetics, perfumery, aromatherapy | |
| 75 | Clary sage / Bahman safed | *Salvia sclarea* | Lamiaceae | Herb | 75%–80% | 0.079%–0.13% | NA | 1.76% | 29 | Linalool, ocimene, nerol,geraniol | Flavouring soft drinks and liquors, beer and wine | |

| 76 | Sandalwood | *Santalum album* | Santala-ceae | Heartwood | 12%–15% | 1.10% | 5% | 1.90% | 72 | Santalol, santalene, curcumene, farnesene | Perfumery, soaps, detergents, shampoo & cosmetics |
|----|-----------|------------------|--------------|-----------|---------|-------|-----|--------|----|--------------------------------------------|----------------------------------------------------|
| 77 | Clove / Lavang | *Syzygium aromaticum* | Myrtaceae | Leaves, stem and bud | 60%–65% | 4.80% | 17.32 | 10.15% | 35 | Eugenol, eugenyl acetate and caryophyllene | Aromatherapy |
| 78 | Marigolds / Zendu | *Tagetes erecta* | Asteraceae | Flower | 70%–95% | 0.65%–7% | NA | | 43 | Tagetone, linalool, limonene, linalyl acetate | Fragrance, cosmetics, pigments as the food colouring |
| 79 | Thyme | *Thymus vulgaris* | Lamiaceae | Herb | 80%–85% | 0.5%–2.5% | | 1.75% | 56 | Thymol, cymene, linalool, limonene, cineole | Aromatherapy, flavouring tooth pastes, candies, ointments, cough syrups |
| 80 | Ajowan/Owa | *Trachysper-mum ammi* | Apiaceae | Seeds | 80%–85% | 1.20% | 1.82% | 2.64% | 12 | Thymol | Flavouring foods, soft drinks, confectionery, soaps and perfumes & medicinal |
| 81 | Vanilla | *Vanilla planifolia* | Orchida-ceae | Seed pods | 25%–30% | 1.2%–2.5% | 4.60% | 7.70% | 60 | Vanillin, eugenol, piperonal, and caproic acid | Flavours, confectionery, ice creams, liquors and in perfumery as well as medicine |
| 82 | Vetiver/Vala/ Khus | *Vetiveria zizanioides* | Poaceae | Roots | 14%–44% | 0.63%–1.5% | 2.41% | 2.23% | 114 | Vetiverol, vetinone, eudesmol & vetiverone | Perfumery & medicinal |
| 83 | Ginger/Ale/ Adrak | *Zingiber officinale* | Zingi-beraceae | Rhizomes | 78%–82% | 1%–3% | 7.48% | 2.62% | 57 | Zingiber-ene, zingiber-one, zingerone, ar-curcumene, farnesene, bisabolene, gingerols & shogaols | Seasoning, blending, natural flavours, pharmaceuticals and perfumery |

## REFERENCES

Busatta, C., Vidal, R. S., Popiolski, A. S., Mossi, A. J., Dariva, C., Rodrigues, M. R. A., Corazza, F. C., Corazza, M. L., Vladimir Oliveira, J., & Cansian, R. L. (2008). Application of Origanum majorana L. essential oil as an antimicrobial agent in sausage. *Food Microbiology*, *25*(1), 207–211. 10.1016/j.fm.2007.07.003

García, D., Alvarez, A., Tornos, P., Fernandez, A., & Sáenz, T. (2002). Gas chromatographic-mass spectrometry study of the essential oils of Pimenta racemosa var. terebinthina and P. racemosa var. grisea. *Zeitschrift Fur Naturforschung - Section C Journal of Biosciences*, *57*(5–6), 449–451. 10.1515/znc-2002-5-608

Kumar, A., Tandon, S., & Yadav, A. (2005). Chemical composition of the essential oil from fresh leaves of Melaleuca leucadendron L. from North India. *Journal of Essential Oil-Bearing Plants*, *8*(1), 19–22. 10.1080/0972060X.2005.10643415

Liu, X. C., Zhou, L. G., Liu, Z. L., & Du, S. S. (2013). Identification of insecticidal constituents of the essential oil of Acorus calamus rhizomes against Liposcelis bostrychophila badonnel. *Molecules*, *18*(5), 5684–5696. 10.3390/molecules18055684

Lodha, A. S., & Telang, A. B. (2016). Cultivation of aromatic plants – A boon for farmers & entrepreneurs in Maharashtra. *Archives of Applied Science Research*, *8*(3), 46–54.

Marongiu, B., Piras, A., Porcedda, S., & Scorciapino, A. (2005). Chemical composition of the essential oil and supercritical CO2 extract of Commiphora myrrha (Nees) Engl. and of Acorus calamus L. *Journal of Agricultural and Food Chemistry*, *53*(20), 7939–7943. 10.1021/jf051100x

Pérez-Alonso, C., Cruz-Olivares, J., Ramírez, A., Román-Guerrero, A., & Vernon-Carter, E. J. (2011). Moisture diffusion in allspice (Pimenta dioica L. Merril) fruits during fluidized bed drying. *Journal of Food Processing and Preservation*, *35*(3), 308–312. 10.1111/j.1745-4549.2009.00457.x

Rahman, M. A., & Islam, M. S. (2015). Alpinia calcarata Roscoe: A potential phytopharmacological source of natural medicine. *Pharmacognosy Reviews*, *9*(17), 55–62. 10.4103/0973-7847.156350

Rout, P. K., Rao, Y. R., Jena, K. S., Sahoo, D., & Mishra, B. C. (2004). Extraction and composition of the essential oil of ambrette (Abelmoschus moschtus) seeds. *Journal of Essential Oil Research*, *16*(1), 35–37).

Sharafan, M., Jafernik, K., Ekiert, H., Kubica, P., Kocjan, R., Blicharska, E., & Szopa, A. (2022). Illicium verum (Star Anise) and Trans-Anethole as valuable raw materials for medicinal and cosmetic applications. *Molecules*, *27*(3), 1–16. 10.3390/molecules27030650

Spadaro, F., Costa, R., Circosta, C., & Occhiuto, F. (2012). Volatile composition and biological activity of key lime citrus aurantifolia essential oil. *Natural Product Communications*, *7*(11), 1523–1526. 10.1177/1934578x1200701128

Thomas, J., Mathew, S., Joy, P. P., & Skaria, B. P. (2016). Aromatic Oils from Kerala. *July*, 2–5. 10.13140/RG.2.1.2832.7924

Unlu, M., Ergene, E., Unlu, G. V., Zeytinoglu, H. S., & Vural, N. (2010). Composition, antimicrobial activity and in vitro cytotoxicity of essential oil from Cinnamomum zeylanicum Blume (Lauraceae). *Food and Chemical Toxicology*, *48*(11), 3274–3280. 10.1016/j.fct.2010.09.001

Wei, F. H., Chen, F. L., & Tan, X. M. (2015). Gas chromatographic-mass spectrometric analysis of essential oil of Jasminum officinale L var grandiflorum flower. *Tropical Journal of Pharmaceutical Research*, *14*(1), 149–152. 10.4314/tjpr.v14i1.21

# Index

For Product Safety Concerns and Information please contact our EU
representative  GPSR@taylorandfrancis.com
Taylor & Francis Verlag GmbH, Kaufingerstraße 24, 80331 München, Germany